Inspire Science

Be a Scientist Notebook

Student Journal

Grade 2

Mc Graw Hill Education

mheducation.com/prek-12

STEM McGraw-Hill is committed to providing instructional materials in Science, Technology, Engineering, and Mathematics (STEM) that give all students a solid foundation, one that prepares them for college and careers in the 21st century.

Send all inquiries to:
McGraw-Hill Education
STEM Learning Solutions Center
8787 Orion Place
Columbus, OH 43240

ISBN: 978-0-07-678222-2
MHID: 0-07-678222-0

Printed in the United States of America.

3 4 5 6 7 8 9 QVS 21 20 19 18 17

Our mission is to provide educational resources that enable students to become the problem solvers of the 21st century and inspire them to explore careers within Science, Technology, Engineering, and Mathematics (STEM) related fields.

TABLE OF CONTENTS

KAYLA
Landscape Architect

TABLE OF CONTENTS

Check out the activities in every lesson!

KAYLA
Landscape Architect

Inspire Science

This is your own journal. You will be a scientist or an engineer. Write in your book as you answer questions and solve problems.

Draw a picture to show what a scientist or an engineer might do.

ERIK
Video Game Designer

 Name ______________________ Date __________

Properties of Matter

Science in My World

Wow, it is hot outside! What do you think of when you see the fruit bars melting in the Sun? Write a question.

____

____

____

____

____

____

Key Vocabulary

Look and listen for these words as you learn about properties of matter.

float	gas	liquid
mass	materials	matter
pattern	property	sink
solid	volume	

Name ______________________ Date ____________

I need to use the best materials when building things.

CHLOE
Carpenter

Chloe wants to be a carpenter. A carpenter cuts and shapes materials to build new things. It is important that carpenters use the correct materials to build all sorts of objects.

Chloe wants to take frozen fruit bars to school, but she doesn't want them to melt. Can you help? Draw what you would use to keep the bars from melting.

Science and Engineering Practices

I will analyze and interpret data.
I will carry out an investigation.
I will plan an investigation.

Name ______________________ Date __________

Describe Matter

What Is Matter?

Circle the things that are matter.

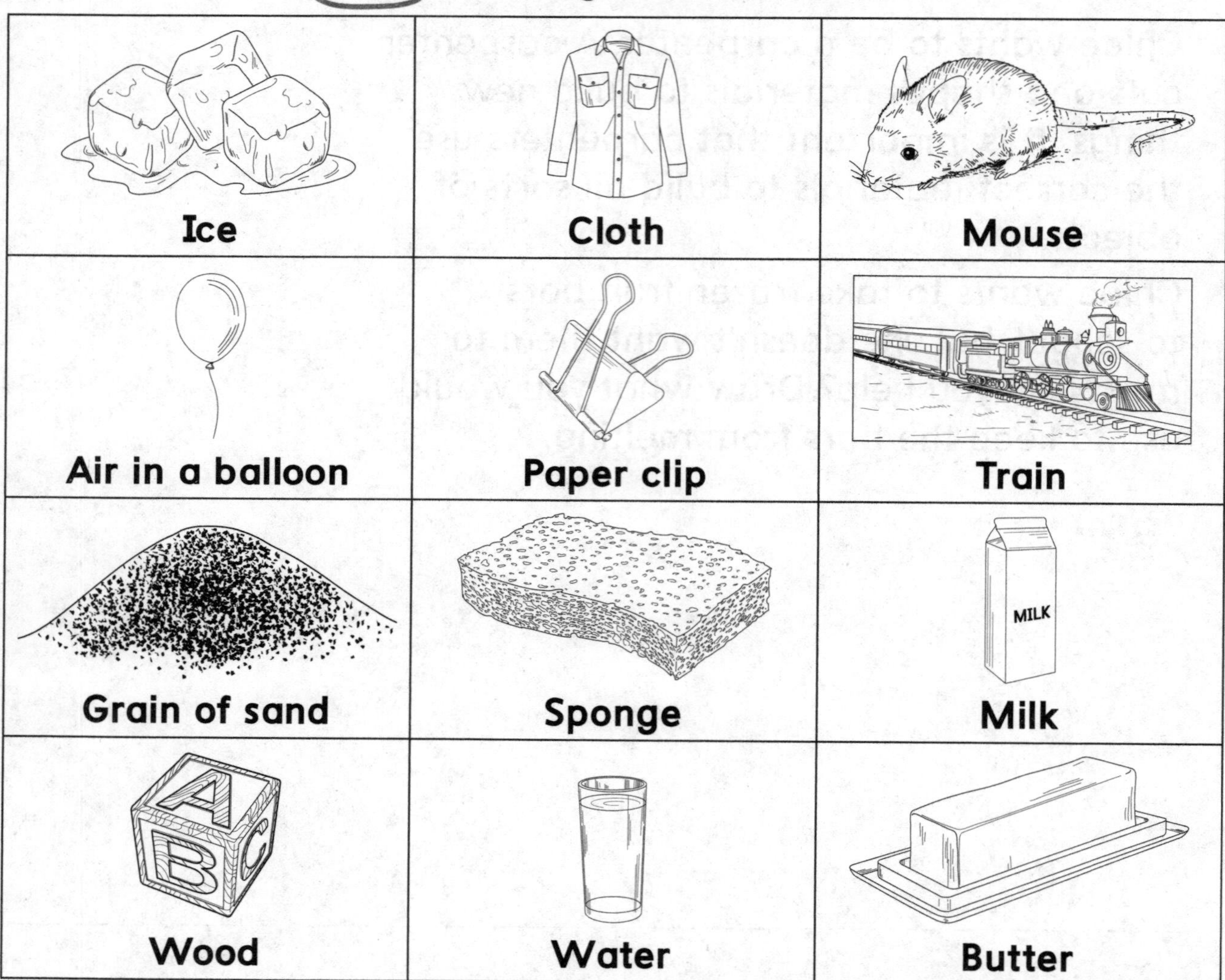

Explain your thinking.

____

____

____

Science in My World

Watch the video of hot air balloons flying high in the sky. How are they able to fly? What questions do you have?

____

____

____

Essential Question

How do we describe matter?

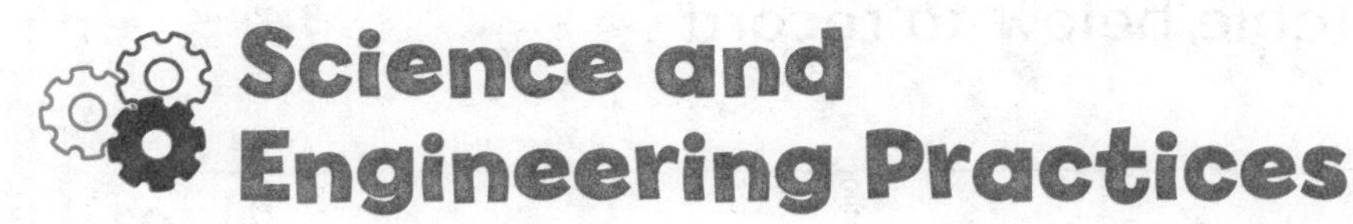

Science and Engineering Practices

I will analyze data.

I love to watch hot air balloons. Are the balloons made of matter?

EMILY
Aerospace Engineer

 Name ______________________ Date __________

Inquiry Activity
Scavenger Hunt

What types of objects can you find around your classoom?

Make a Prediction Which type of matter will you find the most of?

____

Carry Out an Investigation

1 Look around your classroom for different types of matter.

2 Find eight items.

1. **Record Data** Use the table below to record your data.

Solid	Liquid	Gas

Communicate Information

2. **Draw Conclusions** In the table, which type of matter has the most examples? Circle it.

Name ______________________ Date ____________ EXPLAIN

Obtain and Communicate Information

Vocabulary

Use these words when explaining properties of matter.

matter	property	mass
solid	liquid	gas

Matter is All Around Us

Read pages 14-23 in *Matter is All Around Us.* Answer the following questions after you have finished reading.

Fill in the blank.

1. We use our senses to describe the ______________ of matter.

2. Matter is anything that takes up ________.

3. In the table below, draw an example of each type of matter.

 Name ______________________ Date __________

You will sort items from your Scavenger Hunt by their properties.

Make a Prediction Which property will have the most items?

____

____

Carry Out an Investigation

1. Sort the items.
2. Use the table below to record their properties and the names of the items.
3. Circle the items that belong in more than one group.

Properties of Matter			
________	________	________	________

Name ______________________ Date ____________ EXPLAIN

Communicate Information

4. Use the information from your table on page 8 to create a bar graph. Show how many items of each property you sorted.

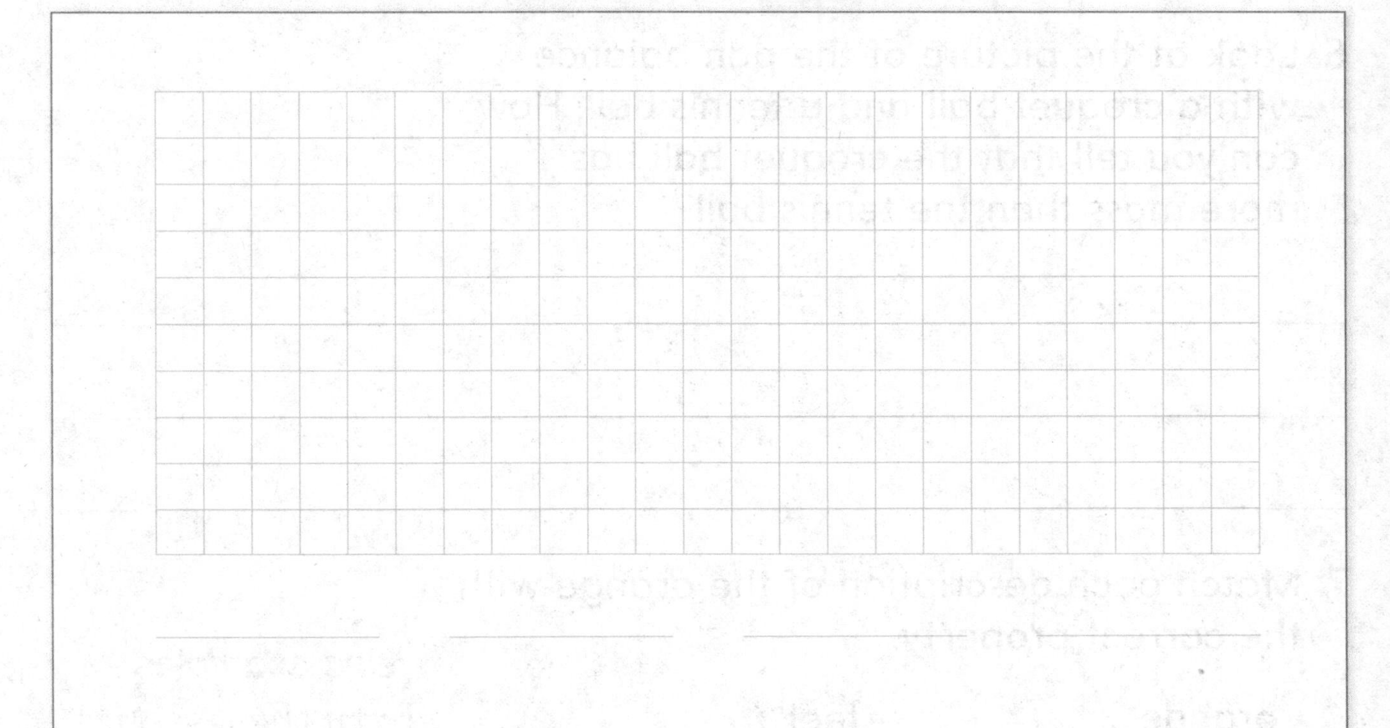

Crosscutting Concepts

Patterns

5. What pattern do you see in the way you sorted or classified matter?

____

____

 Name ______________________ Date __________

What Is Matter?

Read *What Is Matter?* about the properties of matter. Answer the questions after you have finished reading.

6. Look at the picture of the pan balance with a croquet ball and a tennis ball. How can you tell that the croquet ball has more mass than the tennis ball?

____

____

____

7. Match each description of the orange with the correct property.

orange	feel
soft	taste
sweet	look

Use examples from the lesson to explain what you can do!

Science and Engineering Practices

Complete the "I can . . . " statement.

I can analyze data ______________________________

____

____

Research, Investigate, and Communicate

Inquiry Activity

Finding the Mass of Matter

Materials

- [ ] two classroom objects
- [ ] pan balance
- [ ] gram cubes

You will learn how to compare the masses of two objects. Remember, mass is the amount of matter in an object.

Make a Prediction Read the steps of the investigation. Think about the objects you have chosen. Which object do you think will have more mass?

____

____

____

Glue your graph here.

Carry Out an Investigation

1. Look around the classroom for two objects that you can measure with a balance.
2. Choose two objects that you think will have different masses.
3. Measure each object's mass with the gram cubes.
4. **Record Data** Use a separate piece of paper.

 Name ____________________ Date __________

Communicate Information

1. Draw Conclusions How did the masses of your objects compare?

____

____

2. List the two objects on the table. Below each object, list its observable properties. These are properties you can see. Circle any properties that are shared by both objects.

object		
property		

3. How can you sort, or classify, objects using properties?

____

____

4. How would you sort, or classify, objects without seeing them? Explain.

____

____

Name ______________________ Date ____________ EVALUATE

Performance Task

What's in the Bag?

Materials

- ☐ paper bags with items

You can use touch to describe matter. Without looking into a paper bag, you will describe the matter inside.

Make a Prediction How will you describe matter without seeing it?

____

____

Carry Out an Investigation

1. Use the bag provided by your teacher.
2. Reach your hand inside the bag and feel the object. How would you describe what it feels like?
3. **Record Data** Write the bag number and the description in the table.
4. Repeat steps 2 and 3 with other bags.

Bag Number	Description

 Name ______________________ Date __________

Communicate Information

1. **Draw Conclusions** Using the information in your table, can you identify the matter in each bag?

Bag Number	Name of the Object

2. Tell how you were able to identify the matter in the bag.

____

____

____

____

Essential Question

How do we describe matter?

Think about the video of the hot air balloon from the beginning of the lesson. Use what you have learned to describe the matter you observed.

____

____

____

____

____

Science and Engineering Practices

I did analyze data.

Rate Yourself

Color in the number of stars that tell how well you did analyze data.

Now that you're done with the lesson, rate how well you did.

 Name ______________________ Date __________

Solids

Is It a Solid?

Circle the things that are solids.

Rock	Water	Rubber
Feather	Ice	Wool hat
Paper	Juice	Sand
Cotton ball	Air inside a balloon	Nail

Explain your thinking.

____

____

____

Name ______________________ Date __________ ENGAGE

Science in My World

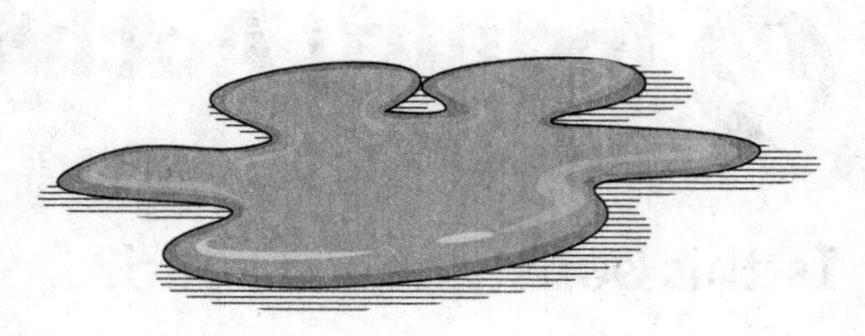

Some matter is hard to describe.
Look at the picture of oobleck.
What questions do you have?

____

____

____

Essential Question

What are the properties of solids?

Like an aerospace engineer, you will conduct investigations to learn about solids.

Science and Engineering Practices

I will carry out an investigation.
I will plan an investigation.

 Name ______________________ Date __________

Inquiry Activity
Oobleck

Is this substance a solid?

Make a Prediction Do you think oobleck is a solid?

____

____

Materials

- ☐ bowl
- ☐ water
- ☐ cornstarch
- ☐ spoon

Carry Out an Investigation

1. Start with some water in the bowl.
2. Add the cornstarch slowly, a little bit at a time.
3. Stir the mixture well until it becomes gooey.
4. **Record Data** Record your results in the table. Then, perform your own actions and record the results.

Is oobleck a solid or liquid?

Action	Result
Squeeze it.	
Make a puddle and quickly drag your fingers through it.	
Roll it into a ball.	
Scoop it with your hand.	

Communicate Information

1. How is oobleck like a solid?

____

____

____

2. What properties of oobleck make it hard to classify?

____

____

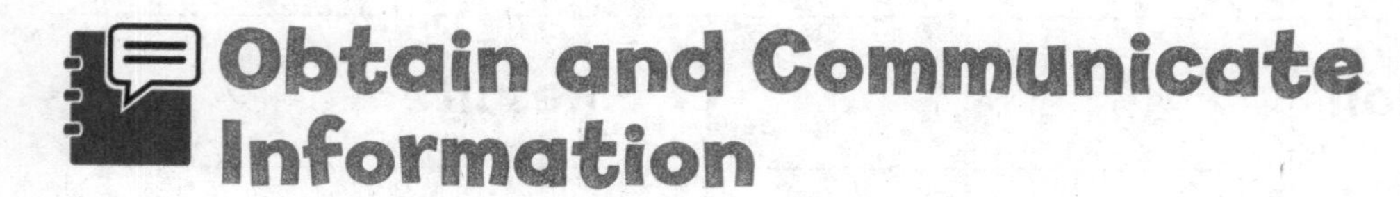

Obtain and Communicate Information

Vocabulary

Use this word when explaining solids.

pattern

1. Match each word with its definition.

property	a state of matter that has a shape of its own
pattern	the repeated way in which something happens
solid	the look, feel, smell, sound, or taste of a thing

2. Circle the matter that is a solid.
Place an X on the matter that is not.

desk	coffee	cracker
air	juice	umbrella

From Nature or From People

Read pages 14-23 in *From Nature or From People.* Answer the following questions after you have finished reading.

3. What do you call materials made by people?

______

4. Name two solids that are found in nature.

______

______

Crosscutting Concepts

Patterns

5. The pattern of a spider web can tell what kind of spider it is. What can other patterns in nature tell us?

______

______

______

 Name ____________________ Date ____________

Inquiry Activity
Identifying Solids

You will plan and conduct an investigation about whether an object is a solid.

Ask a Question Write a question that you would like to answer in the investigation.

____

____

____

____

Carry Out an Investigation

1. Look around the room and choose objects to test.
2. Fill in the "Action" part of the table on the next page to show what you will do to determine if each object is a solid.
3. Use the table to record the name and properties of each object you have chosen.
4. Perform the action on each object.

Name ______________________ Date ____________ EXPLAIN

Communicate Information

5 **Record Data** Fill in the table with your results.

Object	Properties	Action	Result of Action

6. **Draw Conclusions** Circle the objects that are a solid.

7. Look at the objects you circled. How do you know they are solids? Explain your thinking.

____

____

____

Crosscutting Concepts
Patterns

8. What patterns do you see in the solid objects you tested?

____

____

____

What is a Solid?

Read *What is a Solid?* on the properties of solids. Answer the questions after you have finished reading.

9. Circle the words that describe properties.

color	smell	block	taste
size	car	shape	mass

10. Name some materials that solids can be made of.

____

____

____

Name ______________________ Date __________ EXPLAIN

11. Fill in the blank.

To measure a solid's length you can use

a ________________.

To measure the mass of a solid you can use

a ________________.

Science and Engineering Practices

Complete the "I can . . . " statements.

I can carry out an investigation ______________________

____

I can plan an investigation ______________________

____

Use examples from the lesson to explain what you can do!

 Name ______________________ Date __________

Research, Investigate, and Communicate

Inquiry Activity

Measuring Solids

Materials
☐ crayon
☐ book
☐ ruler
☐ pan balance

You will learn how to measure two different solids.

Ask a Question What question will your investigation help you answer?

____

____

Carry Out an Investigation

1. Using a ruler, measure the length of the book and the crayon. Be sure to line up each item carefully with the ruler's edge.
2. **Record Data** Record the measurements in the table on the next page.
3. Using a ruler, measure the width of the book and the crayon. Be sure to line up each item carefully with the ruler's edge. Repeat step 2.
4. Using the pan balance, measure the mass of the book and the crayon. Repeat step 2.

Item	Length	Width	Mass
Book			
Crayon			

Communicate Information

1. Draw Conclusions Did your data support your prediction?

____

____

____

2. How would your data change if you measured two books?

____

____

____

____

____

 Name ____________________ Date ____________

Performance Task

Plan an Investigation about Solids

You will use what you know about solids and oobleck to plan your own investigation.

Ask a Question What question will your investigation help you answer?

____

____

Carry Out an Investigation

1 Write the steps for your investigation.

____

____

____

____

____

____

____

____

____

____

Essential Question

What are the properties of solids?

Think about the picture of oobleck. Use what you have learned to explain if you think oobleck is a solid.

____

____

____

____

Science and Engineering Practices

I did carry out an investigation.

I did plan an investigation.

Now that you're done with the lesson, rate how well you did.

Rate Yourself

Color in the number of stars that tell how well you did plan and carry out an investigation.

 Name ____________________ Date __________

Liquids and Gases

Gases and Liquids

Three friends are talking about matter. Which person has the best idea?

Nate: *Liquids can change to gases. Gases can change back to liquids.*

Arturo: *Liquids can change to gases. Gases cannot change back to liquids.*

Maggie: *Gases can change to liquids. Liquids cannot change back to gases.*

Explain your thinking.

____

____

____

Science in My World

These ice cubes are about to change. Watch the video. What questions do you have?

____

____

____

Essential Question

What are the properties of liquids and gases?

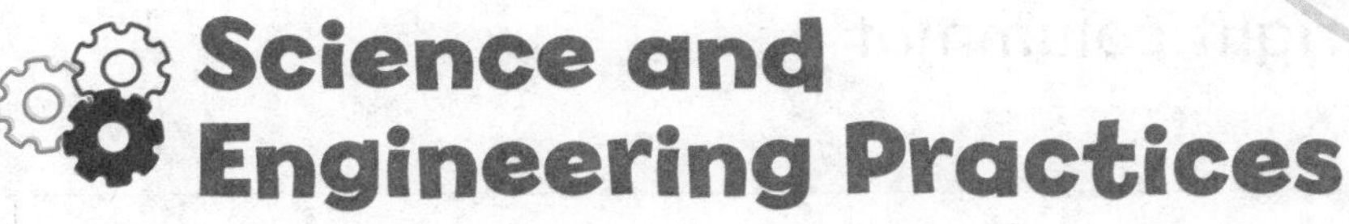

Science and Engineering Practices

I will carry out an investigation.

I will plan an investigation.

States of matter can change. Watch a solid become a liquid. Look even closer to watch a liquid become a gas.

 Name ______________________ Date __________

Inquiry Activity
Measuring Liquids

Materials
- ☐ several containers to hold water
- ☐ tray
- ☐ measuring cup

What happens to the shape of water in different containers?

Make a Prediction How high will the same amount of water be in each of the different containers? Draw your predictions in the left column of the data table.

Carry Out an Investigation

1. Put the containers on the tray.
2. **Record Data** Fill the measuring cup with water. Pour the water into the first container. Draw the container and the water level in the right column of the table below.

My Predictions	My Results

3. Repeat step 2 with the other containers. Use the same amount of water each time.

Communicate Information

1. **Draw Conclusions** Compare your predictions with your results. How are they the same? How are they different?

____

____

____

2. Which container had the highest water level? Why did that happen?

____

____

____

Crosscutting Concepts

Cause and Effect

3. Would the results of the activity change if you used juice instead of water?

____

____

Name ______________________ Date __________

Obtain and Communicate Information

Vocabulary

Use this word when explaining liquids and gases.

volume

What Is a Liquid?

Read *What Is a Liquid?* on the properties of liquids. Answer the questions after you have finished reading.

1. Match the word with its definition.

property	the amount of space something takes up
matter	matter that takes the shape of the container
volume	anything that takes up space and has mass
liquid	the look, feel, smell, sound, or taste of a thing

2. List three liquids in the table below. Then write two words to describe the properties of those liquids.

All About Gas

Explore the Digital Interactive *All About Gas* on the properties of gases. Answer the questions after you have finished reading.

Fill in the blank.

3. A gas has no shape or ______________ of their own.

4. How can you tell that air is made of matter?

____

____

 Name ______________________ Date ____________

Inquiry Activity
Gas Has Mass

Materials

- ☐ balloon
- ☐ tape
- ☐ stick
- ☐ string

You will conduct an investigation to see which balloon has more mass.

Make a Prediction Which balloon do you think will have more mass?

____

Carry Out an Investigation

BE CAREFUL when blowing up your balloon.

1. Blow up your balloon as much or as little as you would like.
2. Compare the size of your balloon to the size of a classmate's balloon.
3. Tape your balloon to the stick balance to compare the mass of your balloon to the mass of a classmate's balloon.

Communicate Information

5. Draw a picture of your balloon and a classmate's balloon on the stick balance.

6. Draw Conclusions How could you tell which balloon had more mass?

____

____

____

Science and Engineering Practices

Complete the "I can . . . " statements.

I can carry out an investigation ______________________________

____

I can plan an investigation ______________________________

____

Use examples from the lesson to explain what you can do!

 Name ______________________ Date __________

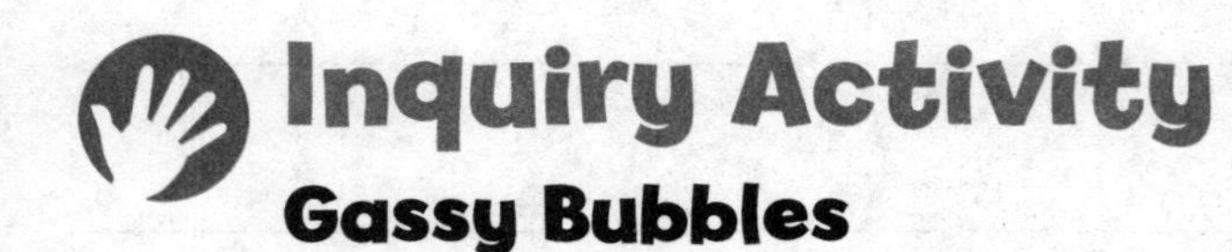

Research, Investigate, and Communicate

Inquiry Activity

Gassy Bubbles

Materials

- ☐ tray
- ☐ water
- ☐ dish soap
- ☐ straw

You will plan and carry out an investigation to see the gas in bubbles.

Ask a Question Write a question that you will investigate.

____

____

Carry Out an Investigation

Plan the investigation by putting the steps in the correct order. Then follow the steps to carry out the investigation.

☐ Using the straw, blow gently into the water.

☐ Add a few drops of dish soap and carefully stir with a straw.

☐ Fill the tray with enough water to cover the bottom.

☐ Record your observations.

Name ______________________ Date ____________

Communicate Information

1. What happened when you blew into the straw? Draw what your tray looked like.

2. **Draw Conclusions** What made the bubbles in the water? What were they filled with?

____

____

____

____

____

Name ______________________ Date __________

Performance Task

All Three States

You will observe an ice cube change from a solid into a liquid, and then into a gas.

Make a Prediction What will happen to the ice cube when it is left in the Sun?

____

____

____

Materials

- [ ] ice cube
- [ ] clear plastic cup
- [ ] plastic wrap
- [ ] rubber band

Glue your graph here.

Carry Out an Investigation

1. Place the ice cube into the plastic cup.
2. Stretch the plastic wrap across the top of the cup, and secure it with a rubber band.
3. Put the cup in a sunny spot in your classroom.
4. Observe changes to the ice cube throughout the day.

Communicate Information

1. Tell what happened to the ice cube throughout the day. Draw your results on a separate sheet of paper.

Essential Question

What are the properties of liquids and gases?

Think about the video of ice cubes on the stove from the beginning of the lesson. Use what you have learned about liquids and gases to explain how the ice cubes changed.

____

____

____

____

Science and Engineering Practices

I did carry out an investigation.

I did plan an investigation.

Rate Yourself

Color in the number of stars that tell how well you did carry out and plan an investigation.

Now that you're done with the lesson, rate how well you did.

Name ______________________ Date __________

Use Matter

Clay Boat

Timmy has a ball of clay. He wants to make a boat out of the clay. Which property should Timmy change to make the clay float?

- ☐ length
- ☐ weight
- ☐ shape
- ☐ color

Explain your thinking.

____

____

____

Name ______________________ Date ____________

Science in My World

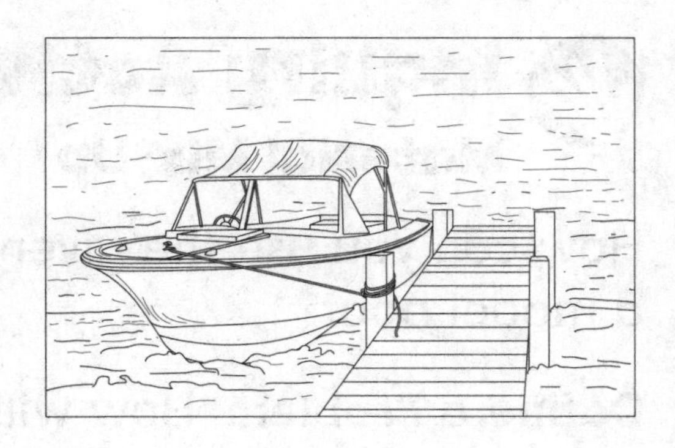

Look at the picture of the boat in the harbor. How does a boat stay on top of the water? What questions do you have?

____

____

____

Essential Question

What materials are best for building a boat?

I love building things! I do a lot of research to find the best materials for what I am building.

Science and Engineering Practices

I will plan and carry out an investigation.

I will analyze and interpret data.

Name ______________________ Date ____________

Inquiry Activity

Material Mix-Up

How can you use the given materials to build a model boat?

Materials

- ☐ paper
- ☐ string or yarn
- ☐ tray
- ☐ water
- ☐ small eraser

Define a Problem How will you use paper and string to build a boat?

____

____

____

____

Make a Model

1. Think about how you can use the materials you have been given to make a boat.
2. Draw a sketch of the boat you will make using these materials.

3 Build your model boat. Test your boat by placing it in a tray filled with water to see if it will float. Place a small eraser on the boat to see if your boat still floats. Draw your observations in the box.

Communicate Information

1. What materials would you use to make your boat better?

____

____

____

2. Why did you choose different materials?

____

____

____

Obtain and Communicate Information

Vocabulary

Use these words when explaining ways to use matter.

materials float sink

1. A heavy rock will ______________ if you throw it into a lake.

2. Builders think about how different ______________ will work in different buildings.

3. If you want your model boat to stay on top of the water, you must use a base that will ______________.

Testing Materials

Investigate how different materials have different properties by conducting the simulation. Answer the questions after you have finished.

4. Which of the materials broke when pulled apart?

____

Crosscutting Concepts
Cause and Effect

5. Use the table below to record how the different materials were affected by different forces.

push	bend
strike	**heat**

6. What happened when the steel was heated? What could you make with steel?

____

____

Matter, Properties, and Making Things

Read pages 14–23 in *Matter, Properties, and Making Things.* Answer the following question after you have finished reading.

7. Circle the statement that is true. Place an X over the statement that is false.

Car designers want strong materials for the body of a car.

When you make or build chairs, you always use hard materials.

Materials Have a Purpose

Watch *Materials Have a Purpose* on how different properties are needed in different products. Answer the questions.

8. Circle the products that you might make with a material that is soft and flexible.

curtains	blanket	skateboard
toy car	table top	chair cushion

9. What was the best material to use for building the bridge? Explain your thinking.

____

____

____

Science and Engineering Practices

Complete the "I can . . . " statements.

I can plan and carry out an investigation ______________________________

____

I can analyze data and interpret data ______________________________

____

____

Use examples from the lesson to explain what you can do!

 Name ______________________ Date __________

Research, Investigate, and Communicate

Boat Research

You will research different types of boats and learn about the materials that help them float.

Research Learn more about materials used to build boats.

Ask a Question What question will your research help you answer?

____

____

____

Record Data Tell what you find.

____

____

Communicate Information

1. **Draw Conclusions** What do you notice about the materials and their properties?

____

____

Performance Task

Make a Model

You will use what you have learned about different materials to plan and build a model boat.

Define a Problem What materials will you use to build a boat that will float?

____

____

Materials

- ☐ assortment of classroom materials
- ☐ clay
- ☐ paper
- ☐ string/yarn
- ☐ tray
- ☐ water
- ☐ small eraser

Make a Model

1. Look around the classroom. What materials will you need? Make a list.

____

____

2. Draw your boat design here. Explain how you will use the materials.

3. Build your model boat.

4 **Test** How will you test whether your boat floats or sinks?

Communicate Information

1. Describe the results of your test.

2. **Draw Conclusions** Compare your boat with others in your class. Why are others floating better? Why are others not floating as well?

3. Draw the boat design that you think worked the best. Label the parts.

Essential Question

What materials are best for building a boat?

Think about the photo of the boat floating in the harbor at the beginning of the lesson. Use what you have learned to describe the materials that are best for building a boat.

______

______

______

______

Science and Engineering Practices

I did plan and carry out an investigation.

I did analyze and interpret data.

Rate Yourself

Color in the number of stars that tell how well you did plan and carry out an investigation and analyze data.

Now that you're done with the lesson, rate how well you did.

 Name ______________________ Date __________

Properties of Matter

Performance Project

Analyze Materials

Materials
- ☐ plastic spoon
- ☐ wooden spoon
- ☐ metal spoon
- ☐ ice cream

You will plan and carry out an investigation to see which spoon is made of the best material.

Make a Prediction Which spoon will scoop hard ice cream the best?

____

____

Carry Out an Investigation

What steps will you use?

____

____

____

____

____

____

____

Which material is best for scooping ice cream?

Communicate Information

Record Data Write or draw your observations.

Explore More in My World

Did you learn the answers to all of your questions from the beginning of the module? If not, how could you design an experiment or do research to help answer them?

 Name ______________________ Date __________

Changes to Matter

Science in My World

What question do you have about the liquid being poured into the mold? Do you notice a change in the liquid? Write a question.

____

____

____

____

____

Key Vocabulary

Look and listen for these words as you learn about changes to matter.

assemble	chemical change	condense
disassemble	dissolve	evaporate
freeze	heat	mixture
physical change	solution	temperature

Name ______________________ Date ____________ MODULE OPENER

How does matter change?

FINN
Construction Manager

Finn wants to be a construction manager. A construction manager plans and makes sure a project stays on schedule. Finn is in charge of a project at the school to pour a new sidewalk that leads to the front door of the school and to the playground. He knows that cool, dry weather is the best for pouring concrete, but how do you think sidewalks get their shape? Draw or write what you think.

Science and Engineering Practices

I will construct explanations.
I will engage in argument from evidence.

 Name ______________________ Date __________

Put Matter Together

Big and Small Blocks

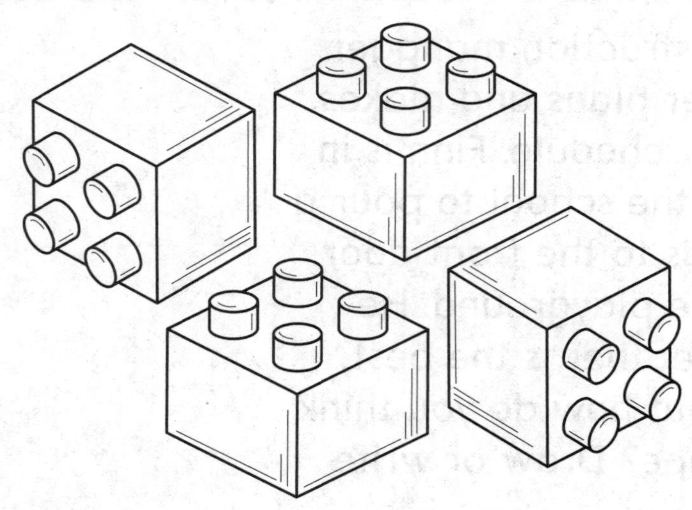

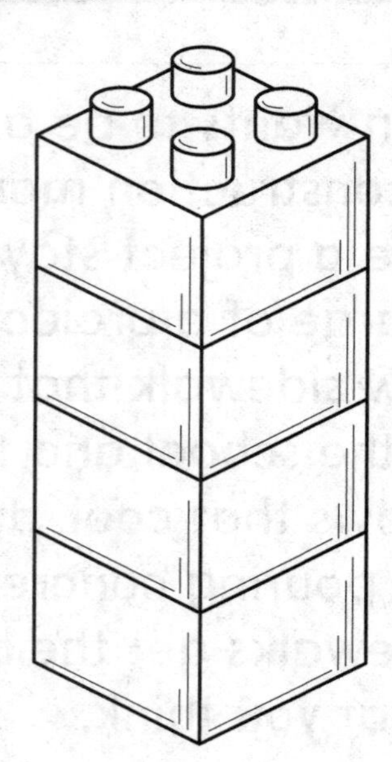

Four small blocks are put together to make one large block. Put an X by the best claim.

- ☐ The large block has more matter than all four small blocks.
- ☐ The large block has less matter than all four small blocks.
- ☐ The large block has the same amount of matter as all four small blocks.

Explain your thinking.

____

____

____

Name ______________________ Date ____________ ENGAGE

Science in My World

Look at the photo of the work site. The workers are using lots of materials. What questions do you have?

____

____

____

____

Essential Question

How can matter be arranged in different ways?

Science and Engineering Practices

I will construct explanations.

I will engage in argument from evidence.

There are lots of different ways that matter can be put together. Let's investigate!
HANNAH
Welder

 Name ______________________ Date __________

Inquiry Activity
Observing Mass with Clay

Materials

- ☐ clay
- ☐ pan balance
- ☐ gram cubes

How does a ball of clay change when you change its shape?

Make a Prediction How will the mass of clay change when you change its shape?

____

____

____

Carry Out an Investigation

1. **Record Data** What shape is the clay at the start of the investigation? Draw it on the left side of the table.
2. Find the mass of the clay using the pan balance and the gram mass set. Record it on the left side of the table.
3. Change the shape of the clay and draw the new shape on the right side of the table.
4. Find the mass of the clay in its new shape. Record it on the right side of the table.

	Start	Finish
Draw the Shape		
Measure the Mass		

Communicate Information

1. **Draw Conclusions** Did the mass of the clay change after you reshaped it?

____

____

____

2. Why or why not?

____

____

____

 Name ______________________ Date __________

Obtain and Communicate Information

Vocabulary

Use these words when explaining changes to matter.

assemble　　physical change　　disassemble

Chef George

Explore the Digital Interactive *Chef George* on physical changes. Answer the questions after you have finished.

1. What different matter did Chef George use to assemble the sandwich?

____

2. How did Chef George change the sandwich after it was assembled?

____

3. How are the pieces of the sandwich alike? How are they different?

____

____

____

Matter Changes

Read *Matter Changes* on the physical changes that can take place in matter. Answer the questions after you have finished reading.

Fill in the blanks.

4. A physical change takes place when the

   size or shape of ____________________
   changes.

5. The ____________________ of matter stays the same if its shape is changed.

6. When something gets wet and then dries,

   it is a ____________________ change.

Inquiry Activity

Get Connected

Materials

- [ ] connecting cubes

You will see how you can make two different objects using the same materials.

Make a Prediction How will the two objects be alike? How will they be different?

____

____

____

Carry Out an Investigation

1. Using a variety of connecting cubes, create an object and draw a picture of it in the box below.

2. Using the same cubes, create a new object and draw a picture of it in the box below.

7. Draw Conclusions Was your prediction correct? How are the two objects alike?

____

____

Crosscutting Concepts

Energy and Matter

8. What did you do to change the shape of the connecting cubes?

____

____

____

Use examples from the lesson to explain what you can do!

Science and Engineering Practices

Complete the "I can . . . " statements.

I can construct explanations ______________________

____

____

I can engage in argument from evidence ______________

____

____

 Name ________________ Date ________

Research, Investigate, and Communicate

Inquiry Activity

Find What Affects Mass

Materials

- ☐ connecting cubes
- ☐ pan balance
- ☐ gram cubes

You will explore the mass of an object when you break it into pieces.

Make a Prediction Will the object's mass be the same once you break it into pieces?

____

____

Carry Out an Investigation

1. **Record Data** Put the object your teacher has given to you on the pan balance. Record its mass in the table.
2. Predict the mass of the pieces. Record your prediction in the table.
3. Take the cubes apart and put all of them on the pan balance. Record their total mass in the table.

Mass of the Object	
Predicted Mass of Pieces	
Actual Mass of Pieces	

Communicate Information

1. **Draw Conclusions** Was your prediction correct? Why or why not?

____

____

____

2. If you made a different shape with the connecting cubes, would it have the same mass as the original object? Tell how you know.

____

____

____

____

____

 Name ______________________ Date __________

Performance Task

Rearranging Matter

Materials

- ☐ clay
- ☐ blocks
- ☐ connecting cubes
- ☐ pan balance

You and a partner will provide evidence that you can make different objects with the same amount and type of matter.

Define a Problem How can you provide evidence that the same amount and type of matter can be used to make different objects?

____

____

____

Carry Out an Investigation

1. Look at the list of the materials you can use to build a new object. Check the box next to the materials that you and your partner will use.
2. **Test** Provide evidence that your object and your partner's object will be made of the same matter.

____

____

____

3. Without looking at your partner's creation, make something with your matter.

4 **Communicate** Describe the object you made, and draw it in the box below.

5 Describe the object your partner made, and draw it in the box below.

6 **Test** Provide evidence that your model and your partner's model are made of the same matter.

____

____

____

Communicate Information

1. How can you use your drawings and tests as evidence that your object and your partner's object are made using the same type and amount of matter but are arranged in different ways?

____

____

____

____

____

____

Name ______________________ Date __________

Essential Question

How can matter be arranged in different ways?

Think about the photo of the work site at the beginning of the lesson. Use what you have learned to tell how the materials can be assembled to build a house.

____

____

____

Science and Engineering Practices

I did construct explanations.

I did engage in argument from evidence.

Rate Yourself

Color in the number of stars that tell how well you did construct explanations and engage in argument from evidence.

Now that you're done with the lesson, rate how well you did.

Name ______________________ Date __________

Mixtures

Mixing Salt and Sand

Anna and her friend mixed some salt and sand together. Which friend has the best idea?

Anna: *I think we can separate the sand from the salt.*

Jerry: *I don't think we can separate the sand from the salt.*

Explain your thinking.

____

____

____

____

____

Science in My World

Watch the video of a mixture being created. Can you see different kind of matter in the bowl? What questions do you have?

____

____

____

____

? Essential Question

What happens when you mix matter together?

Science and Engineering Practices

I will engage in argument from evidence.

I am curious what happens when you mix matter together. Can you separate it? Let's find out!

 Name ______________________ Date __________

Inquiry Activity
Separating Mixtures

How can you separate matter that has been mixed together?

Materials
- ☐ tray
- ☐ Mixture 1-sand and water mixture
- ☐ Mixture 2-rocks and sand mixture
- ☐ Mixture 3-salt and water mixture
- ☐ coffee filter
- ☐ strainer
- ☐ funnel
- ☐ plastic cup

Make a Prediction Which of the mixtures will you be able to separate?

____

____

Carry Out an Investigation

1. With your partner, collect a tray and one of the mixtures your teacher has created.
2. Observe the mixture to decide which filter will work best to separate it.
3. **Record Data** Complete the first table as you attempt to separate your mixture. Record your observations.
4. Repeat the steps to attempt to separate the other two mixtures. Record your observations.

Name ______________________ Date __________ EXPLORE

Mixture Number__________ Parts of Mixture:
What kind of filter did you use? Draw and label your setup.
Record your observations and results.

Mixture Number__________ Parts of Mixture:
What kind of filter did you use? Draw and label your setup.
Record your observations and results.

Mixture Number__________ Parts of Mixture:
What kind of filter did you use? Draw and label your setup.
Record your observations and results.

Name ______________________ Date __________

Communicate Information

1. Which filter best separated Mixture 1?

____

____

2. Which filter best separated Mixture 2?

____

____

3. Which filter best separated Mixture 3?

____

____

4. **Draw Conclusions** How does a filter help you separate mixtures?

____

____

____

____

____

____

____

Name ______________________ Date ____________ EXPLAIN

Obtain and Communicate Information

Vocabulary

Use these words when explaining mixtures.

mixture solution chemical change

dissolve

Mixtures

Watch *Mixtures* on the different types of mixtures. Answer the following questions after you have finished watching.

1. What happens when you have too much water in your sand castle mixture?

____

____

Fill in the blank.

2. Cookies and ______________ are types of food mixtures.

Types of Mixtures

Explore the Digital Interactive *Types of Mixtures* on solid mixtures and solutions. Fill in the table after you have finished.

3. Write or draw your observations.

Type of Mixture	Description	Examples
Solid Mixture		
Solution		

4. Match each word or term with its definition.

chemical change	to mix evenly with liquid and form a solution
solution	two or more different things put together
dissolve	a kind of mixture with parts that do not easily separate
mixture	when matter changes into a different kind of matter

Name ________________ Date ________ EXPLAIN

Methods of Separation

Read *Methods of Separation* on the different ways you can separate mixtures. Answer the questions after you have finished reading.

5. List the different tools you can use to separate mixtures.

____

____

6. Circle the statement if it is true. Place an X over the statement if it is false.

You can use your hands to separate a mixture.

To separate matter that has dissolved in a solution, you can use evaporation.

Science and Engineering Practices

Complete the "I can . . . " statement.

I can engage in argument from evidence.

____

____

____

Use examples from the lesson to explain what you can do!

 Name ______________________ Date __________

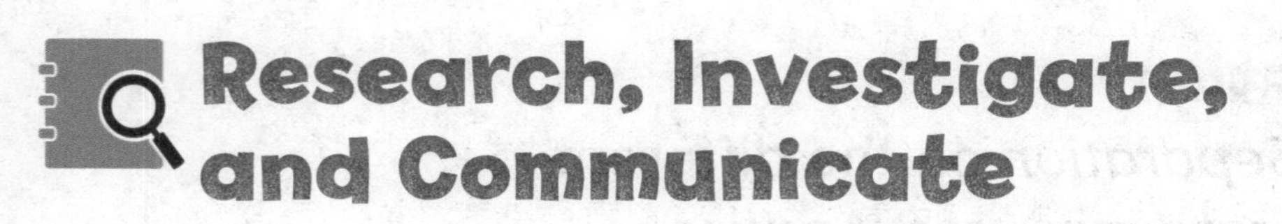

Research, Investigate, and Communicate

Inquiry Activity

Two Liquid Mixtures

Materials

- ☐ glass beaker
- ☐ salt
- ☐ water
- ☐ petri dish
- ☐ eye dropper

You will observe two liquid mixtures. One can be separated, and one cannot.

Make a Prediction How can you separate a mixture of water and salt?

____

Carry Out an Investigation

1. **Record Data** Watch closely as your teacher mixes water and salt together. Draw or write what you observe.

2. Using an eye dropper, put five drops of the salt mixture from your teacher into a petri dish. Draw or write what you observe.

3. Wait one day and observe the petri dish. Draw or write your observations.

4. **Draw Conclusions** What happened when your teacher mixed the water and salt?

____

____

5. How did you separate the solution of water and salt?

____

____

Milk to Stone

Watch *Milk to Stone* on another type of liquid mixture. Answer the following question after you have finished watching.

6. Would you be able to separate this liquid mixture? Why or why not?

____

____

Performance Task

Can You Separate This?

You and a partner will use what you have learned to try to separate a mixture.

Materials

- ☐ mixture of rocks, salt, sand, and water
- ☐ strainer
- ☐ coffee filter
- ☐ funnel
- ☐ tray
- ☐ spoon
- ☐ plastic cup

Make a Prediction Will you be able to separate all of the matter in the mixture?

____

____

Carry Out an Investigation

1. Look at the mixture your teacher has provided. Draw a picture of it in the box below and label the parts.

2. Look at the different items you can use to separate the mixture and decide which matter you should separate first.

3. Fill in the table to complete the investigation.

Name ______________________ Date __________ EVALUATE

Matter to Be Separated	Tool Used to Separate	Write or Draw How You Separated It

Communicate Information

1. **Communicate** What matter did you separate first? Why?

____

____

____

Crosscutting Concepts

Energy and Matter

2. What matter was the most difficult to separate? Why?

3. **Draw Conclusions** Were you able to separate all of the matter from your mixture with the tools given? Why or why not?

Essential Question

What happens when you mix matter together?

Think about the video of the mixture being made. Use what you have learned to explain what happens when the matter in the bowl is mixed together.

____

____

____

____

Science and Engineering Practices

I did engage in argument from evidence.

Rate Yourself

Color in the number of stars that tellhow well you did engage in argument from evidence.

Now that you're done with the lesson, rate how well you did.

 Name ______________________ Date __________

Temperature Changes Matter

Melted Butter

Two friends are melting butter. Which friend has the best claim about the butter?

Lana: *Melted butter can change back to solid butter.*

Tony: *Melted butter can't change back to solid butter.*

Explain your thinking.

____

____

____

____

Name ______________________ Date __________ ENGAGE

Science in My World

Watch the video of the artist heating the glass ball. Why would you heat glass? What questions do you have?

____

____

____

____

Essential Question

How do cooling and heating affect matter?

As a welder, I use heating and cooling to put materials together. I am interested to see other ways heating and cooling change matter!

Science and Engineering Practices

I will engage in argument from evidence.

 Name ______ Date ______

Inquiry Activity
Heat and Ice

Materials

- ☐ 2 cups of water
- ☐ permanent marker
- ☐ thermometer
- ☐ ice

How can heat change ice?

Make a Prediction In which cup will an ice cube melt faster?

______

______

Carry Out an Investigation

1. Label one cup "cold" and one cup "warm." Take the temperature of the cup of cold water.
2. Take the temperature of the cup of warm water.
3. Add an ice cube to each cup.
4. Wait 15 minutes. Observe.
5. Measure the temperature of the water again.

Communicate Information

1. **Record Data** List the temperatures.

	Temperature °C
Cold Water	
Warm Water	
Cold Water after Ice	
Warm Water after Ice	

2. Draw and label your observations.

Before	After an Hour

3. **Draw Conclusions** How did the temperature of the water affect the ice?

____

____

4. How did the temperature of the water change after putting the ice in it?

____

____

 Name ______________________ Date __________

Obtain and Communicate Information

Vocabulary

Use these words when explaining how temperature changes matter.

temperature	thermometer	evaporate
condense	freeze	heat
melt	burn	cool

Matter, Temperature, and Change

Read pages 14-23 in *Matter, Temperature, and Change.* Answer the following questions after you have finished reading.

1. Match each word with its definition.

evaporate	to change from a liquid to a gas and go into the air
condense	to damage and change an object
freeze	change from a solid to a liquid
melt	when a gas changes to a liquid
burn	when an object becomes very cold and changes to a solid

Glue your graph here.

2. Circle the changes that are reversible. Place an X over the changes that are not reversible.

wax melted to make a candle

wood burned in a campfire

lava that came out of a volcano

water that evaporated from a puddle

Change It

Investigate how cooling and heating affect different materials by conducting the simulation. Answer the questions after you have finished.

3. Name one object that goes through a change that is reversible.

____

4. How do you know the change is reversible?

____

____

____

5. Name one object that goes through a change that is not reversible. How do you know?

____

____

Crosscutting Concepts

Cause and Effect

Fill in the blanks.

6. Heating a ____________ changes it into a gas, and cooling a ____________ changes it back to a liquid.

Changes in Matter

Watch the video *Changes in Matter.* Answer the question after you have finished watching.

7. How can a candy bar change states from a liquid to a solid?

____

____

____

Temperature Changes Matter

Read *Temperature Changes Matter* on the ways matter changes with heating and cooling. Answer the question after you have finished reading.

8. What are the clues that a chemical change may be happening?

____

Science and Engineering Practices

Complete the "I can . . . " statement.

I can engage in argument from evidence ______________________

____

____

Use examples from the lesson to explain what you can do!

 Name ______________________ Date ________

Research, Investigate, and Communicate

Inquiry Activity

YOU Change It

Materials

- ☐ variety of objects
- ☐ freezer
- ☐ microwave oven

Now you choose an object to observe as it is cooled and heated.

Carry Out an Investigation

1. Choose one of the objects provided by your teacher.

2. **Ask a Question** What will happen when you cool and heat your object? Will the changes be reversible?

____

____

____

3. What steps will you use to do your investigation?

1. ____

2. ____

3. ____

4. ____

Name ______________________ Date ____________ ELABORATE

Communicate Information

1. Record Data Complete the table.

Object	Effects of Heating	Effects of Cooling

2. Communicate How did your object change when you heated or cooled it?

____

____

____

3. Draw Conclusions Can the change to your object be reversed? Tell how you know.

____

____

____

____

____

Performance Task

Draw the Sequence

Materials

- ☐ paper
- ☐ crayons or colored pencils

You will use what you have learned about cooling and heating matter to draw what happens to butter throughout the day and night.

Make a Prediction What will happen to a stick of butter that was in the sunlight during the day and then left in the kitchen at room temperature overnight?

____

____

Make a Model

1. Fold the paper into thirds.
2. In the first section, draw what the butter looked like when it was put out in the morning.
3. In the middle, draw what the butter looked like in the middle of the day when the sun was shining on it.
4. In the last section, draw what the butter looked like at night when it was room temperature.

Essential Question

How do cooling and heating affect matter?

Think about the video of an artist heating a glass ball at the beginning of the lesson. Use what you have learned to explain how heating and cooling helped shape the glass.

____

____

____

____

Science and Engineering Practices

I did engage in argument from evidence.

Rate Yourself

Color in the number of stars that tell how well you engaged in argument from evidence.

Now that you're done with the lesson, rate how well you did.

Changes to Matter

Performance Project

Design a Solution

Finn was trying to solve the problem of how to shape a sidewalk that leads to the front of the school and to the playground. Using what you have learned about what causes matter to change, help Finn design a mold to shape the sidewalk. Draw your solution below, and list the materials you will need.

Name ______________________ Date ____________

Which materials would be best to create a mold?

What are the steps you would take to create a mold for a sidewalk?

___

___

___

___

___

___

___

___

___

___

Explore More in My World

Did you learn the answers to all of your questions from the beginning of the module? If not, how could you design an experiment to help answer them?

 Name ____________________ Date __________

Earth's Surface

Science in My World

Look at the photo of a polar ice cap. What is an ice cap made of? What other questions do you have?

____

____

____

____

____

Key Vocabulary

Look and listen for these words as you learn about Earth's surface.

continent	fresh water	glacier
globe	island	landform
map	mountain	ocean
river	salt water	stream

Name ____________________ Date __________

What is an ice cap, and where can you find one?

JIN
Paleontologist

Jin wants to be a paleontologist. A paleontologist studies life and fossils from long ago. Jin is looking at the polar ice cap and wondering how long it has been there and what it is made of.

Think about the ways water exists on Earth. Where can we find water? How do we use water? Draw or write what you think.

Science and Engineering Practices

I will develop and use models.

I will obtain, evaluate, and communicate information.

Name ____________________ Date __________

Describe Earth's Surface

Mapping Earth's Surface

Circle the things you might see on a map of Earth's surface.

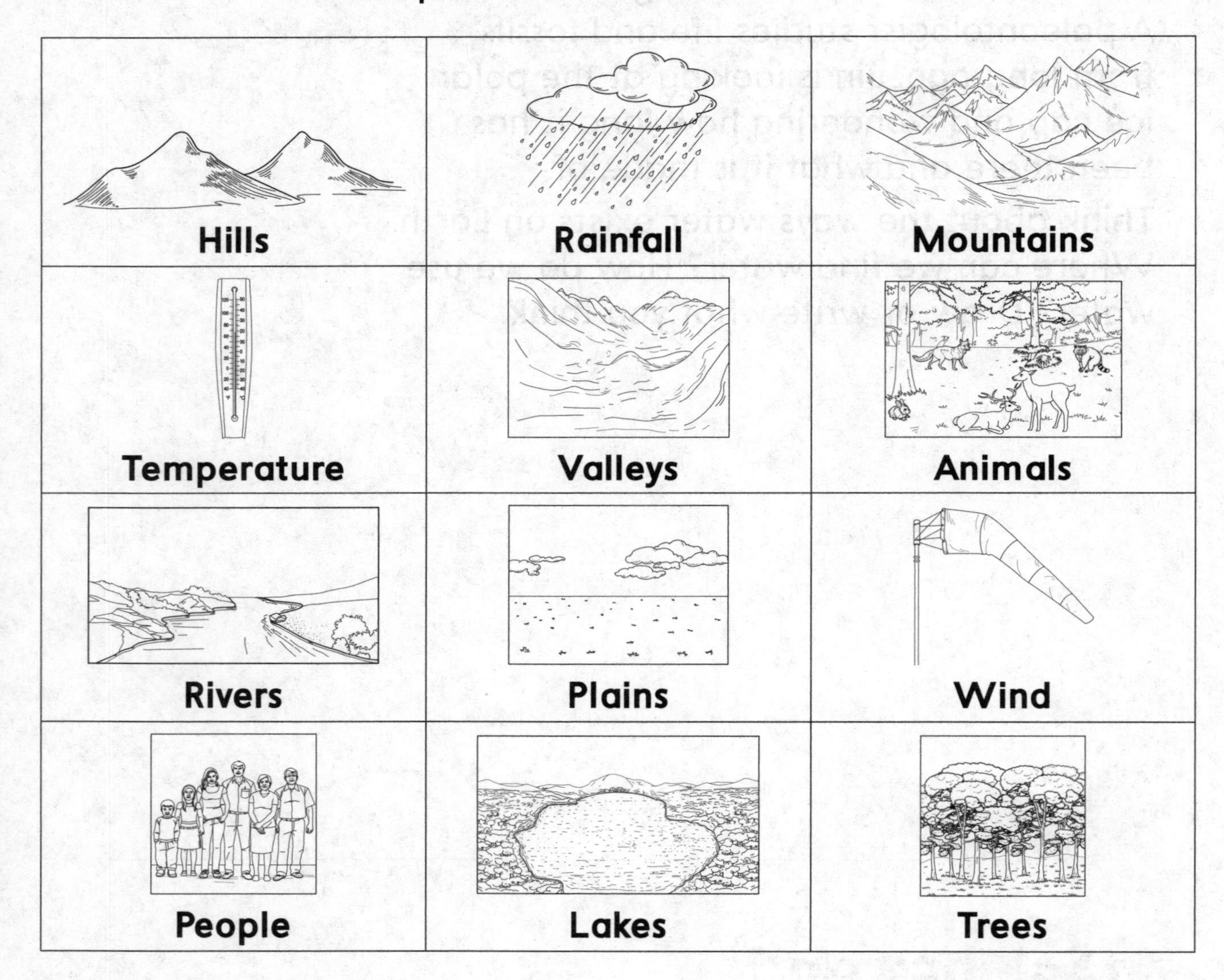

Explain your thinking.

____

____

Science in My World

Look at the photo of the land. What do you notice about how it is shaped? What questions do you have?

____

____

____

____

Essential Question

How can we describe Earth's surface?

Science and Engineering Practices

I will develop and use models.

I will obtain information.

As an ocean engineer I have studied the water around an island, so I'm excited to now investigate the land with you!
HIRO
Ocean Engineer

 Name ______________________ Date __________

Inquiry Activity

Make a Model of Land and Water

Materials
- ☐ modeling clay
- ☐ tray
- ☐ cup of water

How can you show different types of land shapes?

Make a Prediction What can you use to make a model of land and water?

____

____

____

____

____

Make a Model

1. Mold the modeling clay into a land shape you are familiar with.
2. Put the clay land shape into the tray.
3. Pour water around the clay land shape.

Communicate Information

1. **Record Data** Draw and label a picture of your model.

2. How does your model compare to your classmates' models?

___

___

3. Look at your prediction. Were you able to make a model of land and water?

___

___

Obtain and Communicate Information

Vocabulary

Use these words when explaining Earth's surface.

landform	mountain	continent
island	hill	map
plain	valley	

Landforms on Earth

Explore the Digital Interactive *Landforms on Earth* on the different shapes found on Earth's surface. Answer the following questions after you have finished.

1. Match the word with its definition.

continent	an area of land surrounded by water
island	a great area of land on Earth
mountain	land that is very high

2. Name the seven continents on Earth.

____

____

Earth's Surface

Read pages 14–23 in *Earth's Surface*. Answer the following questions after you have finished reading.

Fill in the blank.

3. The many different shapes of Earth's surface are called ______________.

4. Draw pictures showing the landforms you learned about. Label each landform.

 Name ________________________ Date ____________

Watch *Landforms* on ways to describe land. Answer the questions after you have finished watching.

5. Use descriptive words that you learned in the video to add labels to your pictures of landforms on page 107.

____

____

____

6. Circle the statement that is true. Place an X over the statement that is false.

A hill is a raised area of land that is larger than a mountain.

A hill is a raised area of land that is smaller than a mountain.

7. **Research** Pick one of the landforms.

____

8. Describe the landform.

____

____

____

9. Where is that landform found in the world?

____

____

10. List two other facts you learned about the landform.

____

____

11. Draw and label a picture of the landform.

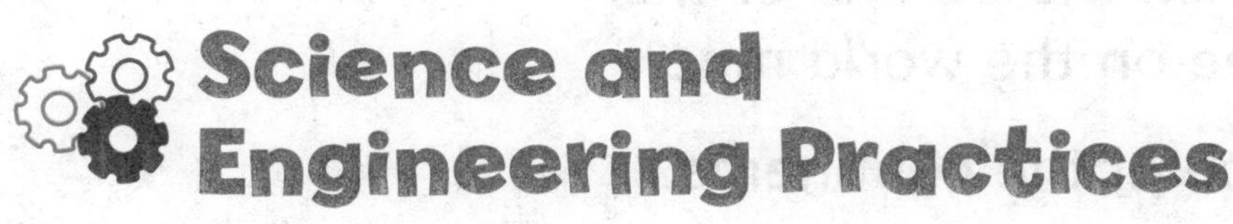

Science and Engineering Practices

Complete the "I can . . . " statements.

I can develop and use models ____________________

____

I can obtain information ____________________

____

Use examples from the lesson to explain what you can do!

Name ________________________ Date ____________

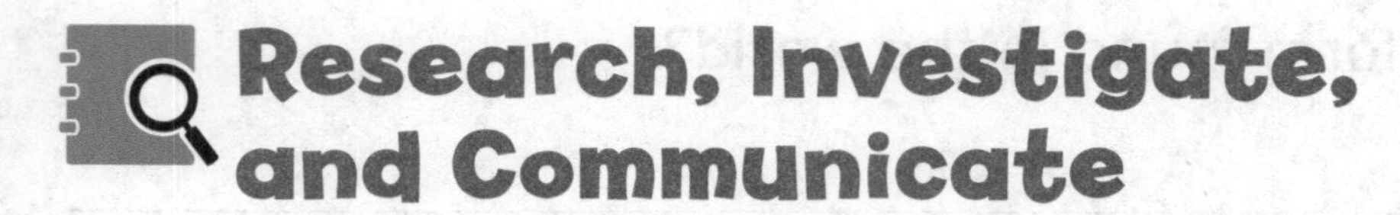

Research, Investigate, and Communicate

Materials
☐ clay
☐ cardboard
☐ world map

Inquiry Activity

Create a Model of a Map

You will create the first layer of a relief map.

Ask a Question What question could you answer using your map?

____

____

Make a Model

1. Look at the world map. Use it to create a relief map, which is a map that shows landforms.
2. Lay the cardboard on a desk or table.
3. On the cardboard, sketch the outline of the continents that you see on the world map.
4. Use clay to create each of the continents, and place them on your map.
5. **Research** Use sources to find two interesting landforms. List them here.

____

____

6. Add them to the map.

Performance Task

Make a Model of a Landform

Materials
☐ clay
☐ cardboard
☐ student chosen materials

You will make a model of a landform.

Ask a Question What question will this investigation help you answer about landforms?

____

____

____

____

Make a Model

1. Lay the cardboard on a desk or table.
2. Mold the modeling clay into a shape that looks like a landform that you have learned about.
3. Add materials to the landform to make it look realistic.
4. Label the landform.

5 Draw your model.

Communicate Information

1. Communicate Describe your landform.

______

______

Crosscutting Concepts

Patterns

2. Compare your model of a landform to those of your classmates who built the same type of landform. What patterns do you notice?

______

______

Essential Question

How can we describe Earth's surface?

Think about the photo of the island at the beginning of the lesson. Use what you have learned to describe Earth's surface.

____

____

____

____

____

Science and Engineering Practices

I did develop and use models.

I did obtain information.

Rate Yourself

Color in the number of stars that tell how well you did develop and use models and obtain information.

Now that you're done with the lesson, rate how well you did.

 Name ______________________ Date ____________

Oceans

Earth's Water

Three friends are wondering where most of Earth's water is found. Which friend has the best idea?

Anna: *I think most of Earth's water is in lakes.*

Lanying: *I think most of Earth's water is in oceans.*

Luis: *I think most of Earth's water is in rivers.*

Explain your thinking.

____

____

____

____

____

Science in My World

Look at the photo of Earth. What do you notice about the different colors and shapes you see? What questions do you have?

____

____

____

____

Essential Question

Where are Earth's oceans?

Science and Engineering Practices

I will develop and use models.

I will obtain, evaluate, and communicate information.

Oceans! I am super excited to help you investigate where Earth's oceans can be found.

 Name ______________________ Date __________

Inquiry Activity
Earth's Surface

How can you use a world map to study Earth's surface?

Make a Prediction What is Earth's surface mostly made of?

____

____

Materials
- ☐ copy of a world map with grid lines
- ☐ green and blue crayons or colored pencils

Carry Out an Investigation

1. Color the squares on the map. If the square is on a continent, or land, color it green.
2. If the square is mostly on the other areas, color it blue.

Communicate Information

1. **Record Data** Count the number of green squares and the number of blue squares. Record your results here.

Number of Green Squares	Number of Blue Squares

2. **Draw Conclusions** Do you have more green squares or blue squares on your map?

3. The green squares show where land is on the map. What do you think the blue squares show?

4. What do your results tell you about the surface of Earth?

 Name ______________________ Date __________

Obtain and Communicate Information

Vocabulary

Use these words when explaining the oceans.

globe   symbol   direction
ocean   salt water

Maps Show Earth's Features

Read *Maps Show Earth's Features* on the different things you can learn from maps. Answer the questions after you have finished reading.

Fill in the blanks.

1. A ____________ is a sphere-shaped map.

2. Maps may use a ___________ to represent the features of an area.

3. North, south, east, and west are examples of ____________.

Where Is Most of Earth's Water?

Explore the Digital Interactive *Where Is Most of Earth's Water?* on Earth's oceans. Answer the question after you have finished.

4. List Earth's five oceans.

____

____

____

Ocean Technology

Watch *Ocean Technology* on how scientists study the ocean floor. Answer the questions after you have finished watching.

5. How does a submersible help scientists study the ocean?

____

____

____

6. Where are the satellites that scientists use to study the ocean?

____

____

Name ______________________ Date __________

Ocean Research

Research You will research where an ocean is located and five facts about the ocean.

Ask a Question What question will your research help to answer?

____

____

Communicate Information

7. **Record Data** Which ocean did you choose to research?

____

8. Where is the ocean located?

____

____

9. List 5 other facts you learned about your ocean.

____

____

____

____

____

Crosscutting Concepts
Patterns

10. What does the ocean you researched have in common with a classmate's ocean?

____

____

Use examples from the lesson to explain what you can do!

Science and Engineering Practices

Complete the "I can . . . " statements.

I can develop and use models ______________________________________

____

____

I can obtain, evaluate, and communicate information ______________

____

____

 Name ______________________ Date __________

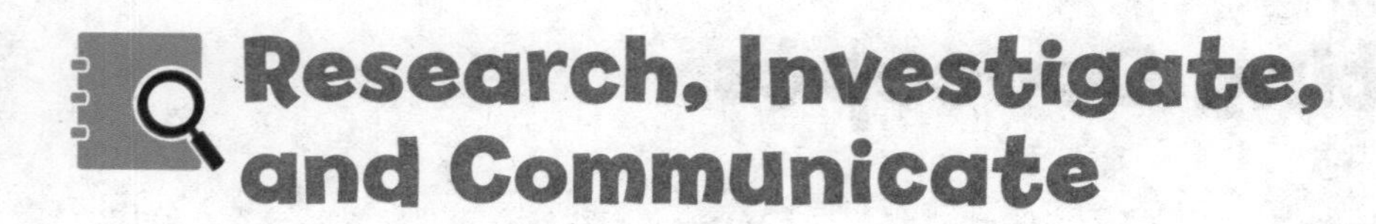

Research, Investigate, and Communicate

Inquiry Activity

Create a Model of a Map

Materials

- ☐ relief map from Lesson 1
- ☐ blue clay
- ☐ world map

You will create the second layer of your relief map.

Ask a Question What question could you answer using your map?

____

____

Make a Model

1. Look at the world map. Use it to locate Earth's oceans.
2. Lay the relief map you created in Lesson 1 on a desk or table.
3. Use the clay to create each of the oceans. Place them on your map and label them.

____

Performance Task

Labeling Earth's Oceans

Materials

- ☐ black-and-white world map on page 124
- ☐ green and blue crayons or colored pencils

You will use a blank world map and label Earth's oceans.

Ask a Question Write a question that you think your map could answer.

____

____

____

Carry Out an Investigation

1. Look at a map of the world and review the names of the oceans.
2. Use crayons or colored pencils on the map to show what is land and what is water.
3. Label the map with the name of each ocean.

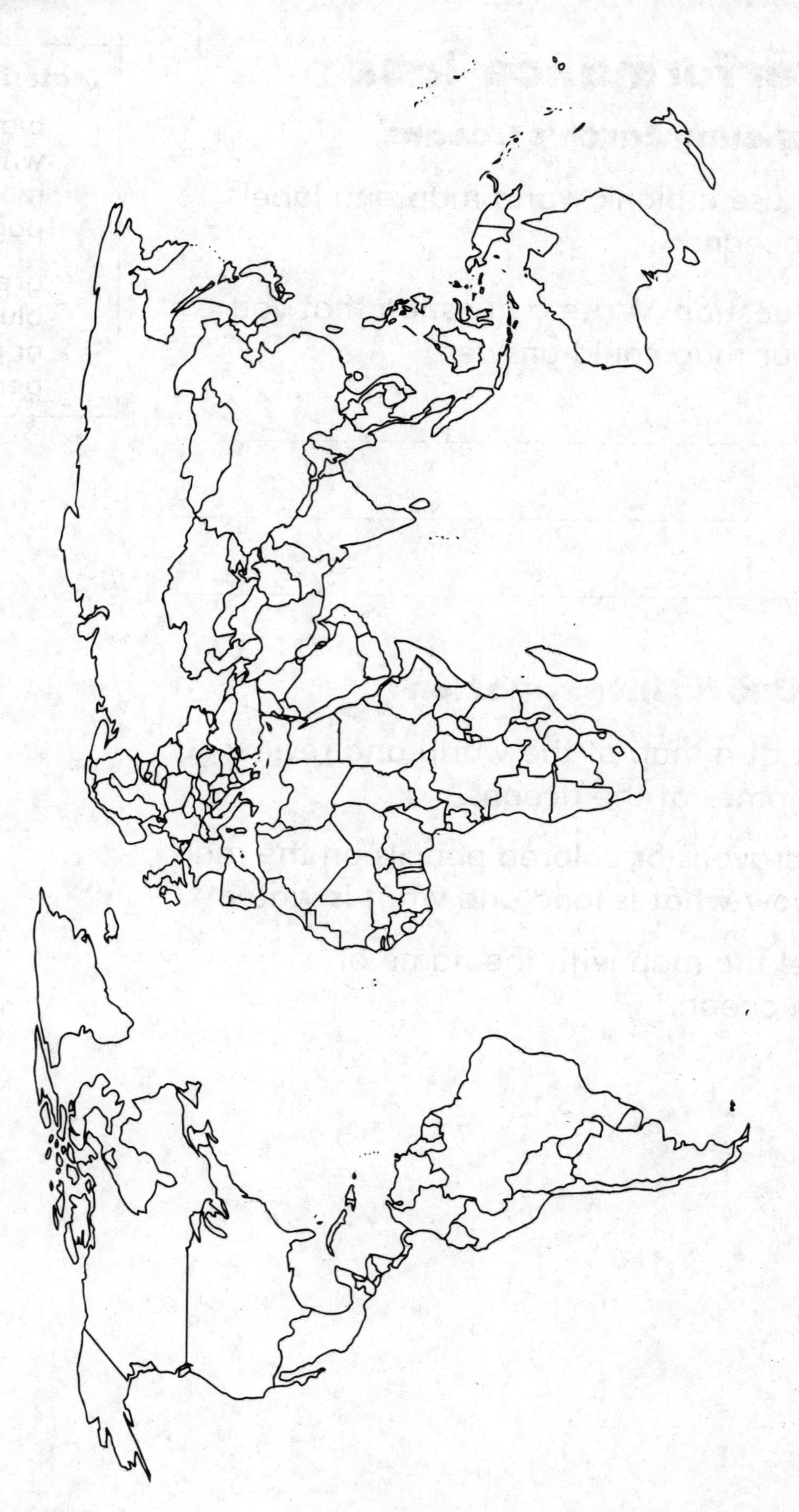

Essential Question

Where are Earth's oceans?

Think about the photo of Earth at the beginning of the lesson. Use what you have learned to describe Earth's surface and where you can find Earth's oceans.

____

____

____

____

Science and Engineering Practices

I did develop and use models.

I did obtain, evaluate, and communicate information.

Rate Yourself

Color in the number of stars that tell how well you did develop and use models and obtain, evaluate, and communicate information.

Now that you're done with the lesson, rate how well you did.

Name ______________________ Date __________

Fresh Water

Fresh Water

Circle all of the places where you can find fresh water.

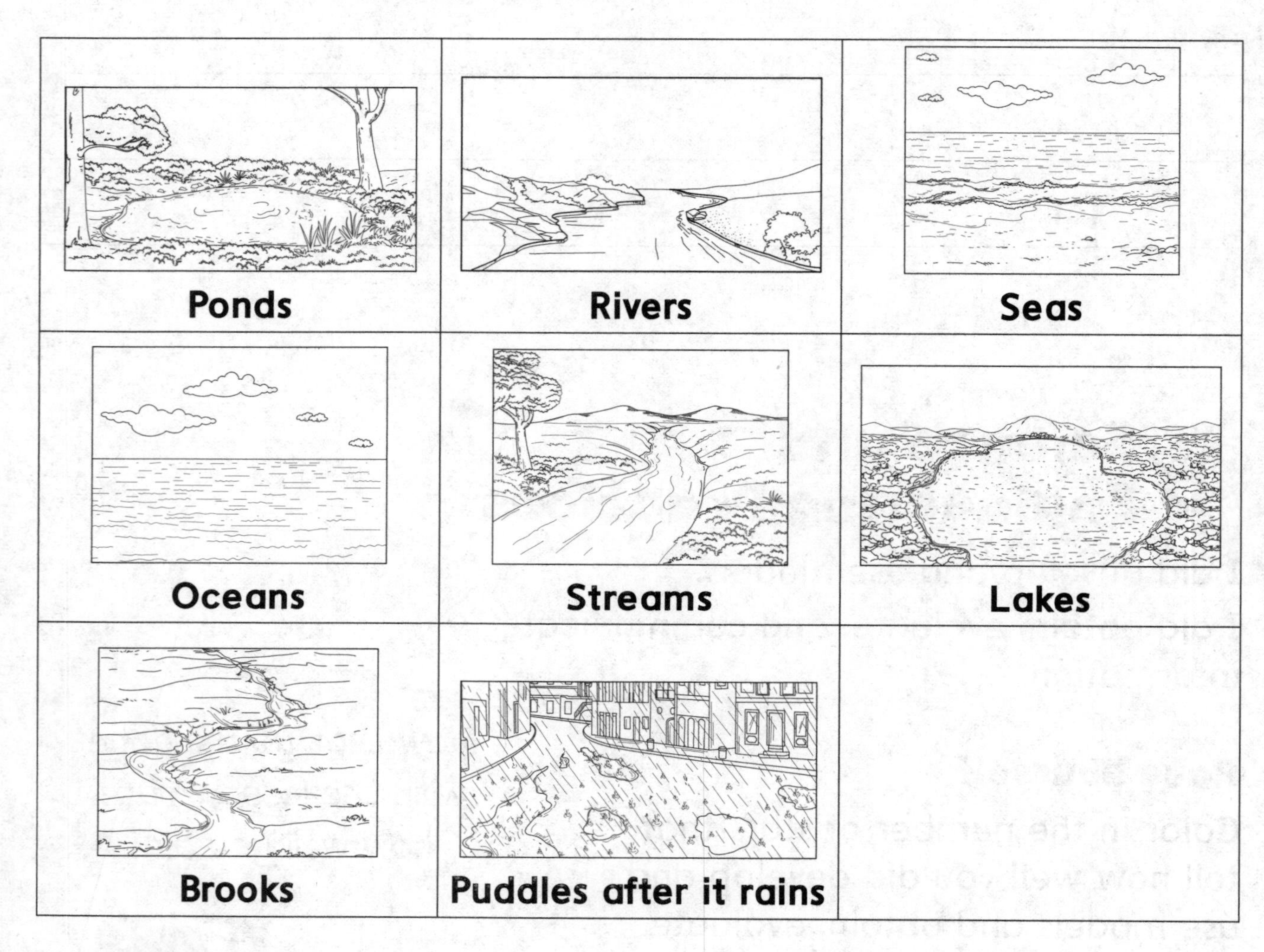

Explain your thinking.

____

____

____

Name ______________________ Date ____________ ENGAGE

Science in My World

Watch the video of the waterfall. Where would you find water like this? What questions do you have?

____

____

____

____

Essential Question

Where is Earth's fresh water?

As an ocean engineer, I study the oceans. I am excited for the chance to explore fresh water on Earth.

Science and Engineering Practices

I will develop and use models.
I will obtain, evaluate, and communicate information.

Inquiry Activity

Fresh Water Changes Simulation

How does fresh water change in different locations and at different temperatures?

Make a Prediction What will happen as fresh water moves down a mountain?

____

____

____

Carry Out an Investigation

Investigate changes in fresh water by conducting the simulation. Answer the questions when you have finished.

1. Observe how the water flows down the mountain when you first open the simulation.
2. Use the slider on the right side to make the mountain scene hotter and colder.

Communicate Information

1. **Record Data** Record your observations in the table.

Hot	Average	Cold
When the temperature in the mountains is 46°	When the temperature in the mountains is 27°	When the temperature in the mountains is 9°

2. Communicate What happens to the water as the temperature gets warmer?

____

____

____

3. Draw Conclusions How does the temperature affect the movement of fresh water?

____

____

____

Obtain and Communicate Information

Vocabulary

Use these words when explaining fresh water.

fresh water	stream	river
lake	pond	glacier

Bodies of Water

Explore the Digital Interactive *Bodies of Water* on sources of fresh water. Answer the question after you have finished.

1. Match each word with its definition.

stream	a small body of fresh water that does not flow
river	a flowing body of water that is smaller than a river
lake	a body of fresh water that has land all around it
pond	a body of water that flows

Water on Earth

Read pages 14–23 in *Water on Earth*. Answer the following questions after you have finished reading.

Fill in the blanks.

2. Oceans have __________ water, while lakes

   and ponds have __________ water.

3. Most of Earth's fresh water is frozen in

   ________________.

4. Name three ways that people use fresh water.

____

____

Fresh Water

Watch *Fresh Water* on the many sources and uses of fresh water. Answer the question after you have finished watching.

5. Which two water sources do not flow?

____

Name ______________________ Date ____________

Inquiry Activity

Make a Model of a Glacier

Materials

- [ ] bowl
- [ ] small rocks
- [ ] shaved ice
- [ ] water
- [ ] freezer
- [ ] tray

You will make a model to see how fresh water is trapped in ice and how it can begin to melt and move.

Make a Prediction What will happen as the glacier begins to melt?

____

____

Make a Model

1. Put a small layer of rocks in the bowl.
2. Put a layer of shaved ice on top of the rocks.
3. Repeat steps 1 and 2 until the bowl is almost full.
4. Carefully pour a small amount of water in the bowl, but do not fill it to the top.
5. Put the bowl in the freezer until the water freezes.
6. Take the model out of the bowl and place it upside down inside the tray.

Glue your graph here.

Name ______________________ Date __________

Communicate Information

6. Record Data What happened as the glacier started to melt?

____

____

7. Push the model with your finger. What happens?

____

____

8. Draw Conclusions What happens if a glacier doesn't melt?

____

____

Use examples from the lesson to explain what you can do!

Science and Engineering Practices

Complete the "I can . . . " statements.

I can develop and use models ______________________

____

I can obtain, evaluate, and communicate information ______________________

____

 Name ____________________ Date __________

Research, Investigate, and Communicate

Fresh Water Research

Research You will research a lake or river, where it is located, and five facts about it.

Ask a Question What question will your research help to answer?

____

____

Communicate Information

1. **Record Data** Which lake or river did you choose to research?

____

____

2. Where is the lake or river located?

____

____

3. List five other facts you learned about your lake or river.

____

____

____

____

4. Do parts of your lake or river ever become solid? Does it ever have glaciers?

____

____

Crosscutting Concepts

Patterns

5. Compare your research with that of a classmate who researched a different lake or river. What patterns do you see?

____

____

 Name ______________________ Date __________

Inquiry Activity
Create a Model of a Map

Materials
- relief map from Lessons 1 and 2
- blue clay

You will create the final layer of your relief map.

Ask a Question What question could you answer using your map?

___

___

Make a Model

1. Look at your relief map. Find the approximate location of the lake or river that you researched.
2. Create the body of water and place it on your map. Label it.
3. Ask four classmates for the name and approximate location of their lake or river to place on your map.

___

___

___

4. Use the clay to create the additional bodies of water. Place them on your map and label them.

Performance Task

Make a Model of Fresh Water Movement

Materials

- [ ] crayons or colored pencils

You will draw a model that shows how fresh water moves down or around a landform.

Make a Prediction How can you draw a model of fresh water movement?

____

____

____

____

Make a Model

1. Revisit the simulation *Fresh Water Changes*.

   Investigate Fresh Water Changes by conducting the simulation.

2. Watch the water as it moves down the mountain.

Communicate Information

Record Data Draw an outdoor model that includes at least one landform and three sources of fresh water. Label the sources of fresh water. Draw arrows to show how the water flows in your outdoor model.

? Essential Question

Where is Earth's fresh water?

Think about the video of the waterfall at the beginning of the lesson. Use what you have learned to explain where Earth's fresh water is.

____

____

____

____

Science and Engineering Practices

I did develop and use models.

I did obtain, evaluate, and communciate information.

Rate Yourself

Color in the number of stars that tell how well you did develop and use models and obtain, evaluate, and communicate information.

Now that you're done with the lesson, rate how well you did.

Earth's Surface

Performance Project

Polar Ice Cap Research

Research You will research polar ice caps and where to find solid ice on Earth.

Ask a Question What question will your research help to answer?

___

___

___

Communicate Information

1. **Record Data** Where can you find polar ice caps and other solid ice on Earth?

___

___

___

Let's research where we can find solid ice on Earth.

2. What else did you learn about polar ice caps?

Explore More in My World

Did you learn the answers to all of your questions from the beginning of the module? If not, how could you design an experiment to help answer them?

 Name ______________________ Date __________

Earth's Surface Changes

Science in My World

Look at the photo of the flooding. Flooding can happen when it rains a lot in a short period of time or when a mass of ice melts. What questions do you have about flooding?

____

____

____

____

____

abc Key Vocabulary

Look and listen for these words as you learn about changes to Earth's surface.

coast	earthquake	erosion
flood	landslide	natural resource
rock	sand	soil
volcano	weathering	windbreak

Name ______________________ Date ____________

How can the surface of Earth change?

Maya wants to be a geologist. A geologist studies Earth, what it is made of, and how it has changed. While on a hike along a stream, Maya noticed her favorite wildflower field was now under water. What do you think happened? Draw a picture of how you think the area would look.

Science and Engineering Practices

I will construct explanations.
I will design solutions.

Name ______________________ Date __________

Weathering and Erosion

Shapes of Landforms

Three friends are talking about changes to the shapes of landforms. Which friend has the best idea?

Carlos: *I think water can change the shape of land.*

Penny: *I think wind can change the shape of land.*

Max: *I think both water and wind can change the shape of land.*

Explain your thinking.

____

____

____

Science in My World

Look at the photo of a rock in the water. What do you notice about how the rock is shaped? What questions do you have?

____

____

____

____

Essential Question

How can wind and water change Earth's surface?

I am curious about how wind, water, and weather affect Earth's surface.
HUGO
Meteorologist

Science and Engineering Practices

I will construct explanations.

 Name ______________________ Date ____________

Inquiry Activity
How Can You Change Rocks?

What do you think will happen when chalk is shaken together with pebbles and water?

Materials
- ☐ 10 pebbles
- ☐ 8 small chalk pieces
- ☐ jar with lid
- ☐ hand lens
- ☐ water

Make a Prediction What will happen to pieces of chalk when they are shaken in a jar with pebbles?

____

____

Carry Out an Investigation

1. Place 4 pieces of chalk and 10 pebbles in a jar. Shake the jar for two minutes.
2. **Record Data** Draw or write about the chalk.

3. Empty the jar and fill it half full with water. Add 10 pebbles and 4 new pieces of chalk.

4. Shake the jar for two minutes. Draw or write about the chalk.

Communicate Information

1. **Draw Conclusions** How did the chalk change after being shaken in water with rocks?

___

___

2. Why do you think the chalk changed more after being shaken in water?

___

___

___

Name ______________________ Date __________

Obtain and Communicate Information

Vocabulary

Use these words when explaining weathering and erosion.

erosion	rock	soil
flood	weathering	sand

Our Changing Earth

Read pages 14–23 in *Our Changing Earth* on how weathering and erosion affect Earth's surface. Answer the questions after you have finished.

Fill in the blanks.

1. Erosion happens when ____________ and ____________ are moved by wind or water.

2. Strong winds in the desert blow ____________ and can cause weathering to the land.

3. As it melts, a ____________ moves slowly across the land and breaks off pieces of rock.

Name ______________________ Date ____________ EXPLAIN

Inquiry Activity
Erosion

You will observe the effects of wind erosion on a mountain of sand.

Materials

- [ ] safety goggles
- [ ] plastic cup
- [ ] sand
- [ ] tray
- [ ] straw

Make a Prediction What will happen to a sand mountain when you slowly blow on it?

____

____

Carry Out an Investigation

BE CAREFUL Wear safety goggles.

1. Pack sand in the plastic cup.
2. Place the cup upside down on your tray. Remove the cup from the sand.
3. **Record Data** Draw a picture of the sand mountain.

 Name ______________________ Date ____________

4 Use the straw to blow **gently** one time on the side of the mountain. Blow gently two more times and observe the changes.

5 Draw a picture to show how the mountain and tray look now.

6 Using the straw, blow 10 more times on the mountain. You can blow on the top or the sides. Blow until you see changes to the shape and size of the mountain.

7 Draw a picture of what the mountain looks like now.

Communicate Information

4. **Draw Conclusions** How did the sand mountain change after the first time you blew on it?

5. How did the mountain change after you blew on it 10 times?

____

____

____

____

6. Where did the sand go when it came off the mountain?

____

____

____

____

Crosscutting Concepts

Stability and Change

7. Did the sand mountain change slowly or rapidly?

____

____

____

____

How Landforms Are Made

Explore the Digital Interactive *How Landforms Are Made* on how different landforms have been shaped after years of weathering and erosion. Answer the question after you have finished.

8. Match each landform to the type of erosion that helped form it.

sea arch	wind erosion
sand dune	wind and water erosion
canyon	water erosion
rock pedestal	

Use examples from the lesson to explain what you can do!

Science and Engineering Practices

Complete the "I can . . . " statement.

I can construct explanations ______________________

____

____

____

 Name ______________________ Date __________

Research, Investigate, and Communicate

Inquiry Activity

Model Weathering

Materials

- ☐ sugar cube
- ☐ tray
- ☐ pipette
- ☐ cup of blue water

You will observe how water can slowly break down an object.

Make a Prediction How many drops of water will it take for the water to change the shape of a sugar cube?

____

Carry Out an Investigation

1. Put the sugar cube in the middle of the tray.
2. Use the pipette to squeeze 5 drops of water onto the sugar cube. Look closely for any changes.
3. Continue to squeeze drops of water onto the sugar cube until you can see a change in the shape of the cube.
4. **Record Data** Use tally marks to show how many drops you put on the cube before you saw a change in its shape.

Tally Marks

5 Continue to drop water onto the sugar cube until it is gone. Use tally marks to show how many more drops it took.

Tally Marks

Communicate Information

1. **Draw Conclusions** How many drops of water did it take to change the shape of the sugar cube?

____

2. How many drops of water did it take until the sugar cube was gone?

____

3. Where did the sugar from the cube go?

____

____

4. Do you think the sugar cube underwent a fast change or a slow change? Tell how you know.

____

____

EVALUATE Name ______________________ Date __________

Performance Task

Earth's Slow Changes

You will use what you have learned to draw an example of weathering or erosion that could happen near your home.

Make a Prediction How can you draw a model of a slow change?

____

____

Make a Model

1. Plan your model by answering the questions below.

Communicate Information

1. **Record Data** Which landform did you choose?

____

2. Will your drawing show erosion or weathering?

____

2. On a separate piece of paper, draw a model of the landform. Label the parts of the landform where weathering or erosion have happened.

Glue your model here.

Name ______________________ Date __________

Essential Question

How can wind and water change Earth's surface?

Look back at the photo of the sea arch from the beginning of the lesson. Use what you have learned to tell how that rock got its shape.

____

____

____

____

____

Science and Engineering Practices

I did construct explanations.

Rate Yourself

Color in the number of stars that tell how well you did construct explanations.

Now that you're done with the lesson, rate how well you did.

 Name ______________________ Date __________

Quick Changes to Earth's Surface

Quick Changes

Which things can cause very quick changes to Earth's surface? Circle them.

Volcanoes   Weathering   Floods

Landslides   Earthquakes   Erosion

Explain your thinking.

____

____

____

Name ______________________ Date ____________

Science in My World

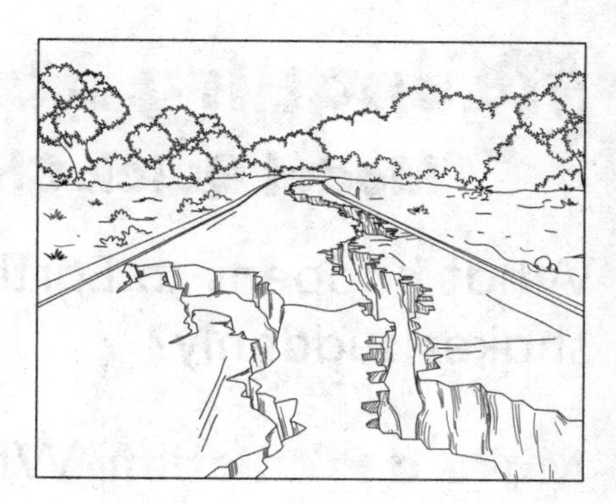

Look at the photo of the damaged road. Some changes on Earth happen quickly. What questions do you have?

____

____

____

____

Essential Question

How can Earth's surface change quickly?

Science and Engineering Practices

I will construct explanations.

I study all kinds of natural events. Did you know there are things that can cause BIG changes to Earth?

 Name ______________________ Date __________

Inquiry Activity

Model Quick Changes to Earth

Materials

- tray
- sand
- assorted blocks
- twigs

What happens to Earth when it moves or shakes suddenly?

Make a Prediction What will happen to an outdoor model with trees and buildings when it suddenly shakes?

____

____

Make a Model

1. Fill the tray halfway with sand. Form a mountain of sand somewhere in the tray.
2. Use the blocks to create buildings. Stick twigs in the sand to model trees.
3. **Record Data** In the box below, draw a picture of your outdoor model. Label Earth's surface, the trees, and the buildings.

4 **Test** Tap or twist the tray gently once or twice. Record your observations.

---

5 Now tap or twist the tray a bit harder. Record your observations.

---

6 In the box below, draw a picture of your outdoor model now.

 Name ____________________ Date __________

Communicate Information

1. **Draw Conclusions** How did the quick tapping and twisting change your outdoor model?

____

____

____

____

____

2. Changes can happen quickly or slowly. Describe these changes.

____

____

____

____

____

Name ______________________ Date __________ EXPLAIN

Obtain and Communicate Information

Vocabulary

Use these words when explaining quick changes to Earth's surface.

landslide earthquake volcano
erupt lava

Events That Change Earth's Surface

Read pages 14–23 in *Events That Change Earth's Surface* on how Earth's surface can change in different ways. Answer the questions after you have finished.

1. Fill in the blanks.

   An ______________________ may last only 30 seconds, but a ______________________ may erupt for many days.

2. How does weathering cause changes to Earth's surface?

____

____

____

Parts of a Volcano

Explore the Digital Interactive *Parts of a Volcano* on what happens as a volcano erupts. Answer the question after you have finished.

3. Circle the statement if it is true. Place an X on the statement if it is false.

The crater of a volcano holds the lava.

When a volcano erupts, lava flows down the flanks.

Inquiry Activity

Volcano Eruption

You will observe how a model of a volcanic eruption changes Earth's surface.

Make a Prediction What will happen to the sides and surrounding Earth of a volcano when it erupts?

____

____

____

Materials

- ☐ safety goggles
- ☐ tray
- ☐ small jar
- ☐ clay
- ☐ dishwashing liquid
- ☐ spoon
- ☐ baking soda
- ☐ vinegar

Make a Model

BE CAREFUL Wear goggles to protect your eyes.

1. Center the jar in the tray. Shape the clay around the jar to make it look like a volcano.
2. Squeeze 2–3 drops of dishwashing liquid into the jar.
3. Fill the jar halfway with vinegar.
4. Add a spoonful of baking soda to the jar.
5. **Record Data** In the box below, draw or write your observations.

Communicate Information

4. Communicate What happened when your model erupted?

____

____

 Name ____________________ Date __________

Crosscutting Concepts
Stability and Change

5. Draw Conclusions How did the quick eruption change the volcano and the land around it?

____

____

Earthquakes

Read *Earthquakes* on the changes caused by a historic earthquake. Answer the question after you have finished reading.

6. The earthquake caused damage to

______________, ______________,

and ______________.

Use examples from the lesson to explain what you can do!

Science and Engineering Practices

Complete the "I can . . . " statement.

I can construct explanations ______________

____

____

Research, Investigate, and Communicate

Landslides

Explore the Digital Interactive *Landslides* on how a landslide can quickly change Earth. Answer the question after you have finished.

1. What are three causes of landslides?

____

____

Flooding Research

Research Look through the materials provided by your teacher to learn more about floods.

Ask a Question What question will your research help to answer?

____

____

Communicate Information

Record Data Fill in the chart with the events that are a part of flooding.

 Name ______________________ Date __________

Communicate Information

Record Data Fill in the events that happen when an area floods.

First

Next

Then

Finally

Name ______________________ Date ____________ EVALUATE

Performance Task

Make a Model of a Quick Change

Materials

- [ ] crayons or colored pencils

You will draw a before-and-after picture that shows a quick change to Earth's surface.

Make a Prediction How can you draw a model of a quick change?

____

____

____

____

Make a Model

Recall what you have learned about the quick changes to Earth's surface. Circle the change you would like to show in your before-and-after picture.

Earthquake **Volcano** **Landslide** **Flood**

Communicate Information

1. **Record Data** Draw what the outdoor model looks like before the quick change happens. Label the landforms and Earth's surface.

2. Now draw what the same outdoor model looks like after the quick change. Label the landforms, Earth's surface, and the changes that happened.

Essential Question

How can Earth's surface change quickly?

Think about the photo of a damaged road from the beginning of the lesson. Use what you have learned to tell why the road may look that way.

____

____

____

____

____

Science and Engineering Practices

I did construct explanations.

Rate Yourself

Color in the number of stars that tell how well you did construct explanations.

Now that you're done with the lesson, rate how well you did.

 Name ______________________ Date __________

Slowing Earth's Changes

How Do People Change Earth's Surface?

Three friends are talking about how people change the surface of Earth. Which friend has the best idea?

Frankie: *I think people change Earth's surface in a good way.*

Casey: *I think people change Earth's surface in a bad way.*

Holly: *I think people change Earth's surface in both good and bad ways.*

Explain your thinking.

____

____

Name ______________________ Date __________ ENGAGE

Science in My World

Look at the photo of a sandy beach. What do you notice about the piles of sand? What questions do you have?

____

____

____

Essential Question

How can people slow the changes to Earth's surface?

Science and Engineering Practices

I will construct explanations.

I will design solutions.

I study the way weather changes Earth. Did you know there are things YOU can do to help slow down weathering and erosion?

Inquiry Activity

Beach Erosion

Materials

- [ ] tray
- [ ] sand
- [ ] water
- [ ] empty water bottle

What happens to a beach when water and waves touch the sand?

Make a Prediction What will you observe when you make a model of the beach and waves?

____

____

Make a Model

1. Pour sand into one end of the tray and build it up to create a beach.
2. Carefully fill the other end of the tray with enough water to touch the sand.
3. **Record Data** Draw a picture of your beach model in the box below.

4 Lay a bottle across the tray in the water. Gently push it down into the water with your fingertips to create waves for two minutes.

5 Observe what happens to the sand on the beach.

6 Draw a picture of your beach model now.

Communicate Information

1. What happened to the sand on the beach?

____

____

2. How could you design a solution to keep the sand from washing away?

____

____

Name ______________________ Date __________

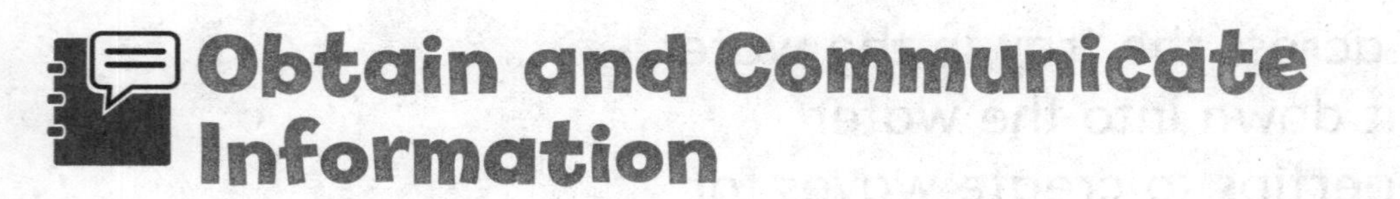

Obtain and Communicate Information

Vocabulary

Use these words when explaining how people can help slow the changes to Earth's surface.

coast windbreak natural resource

Coastal Erosion

Watch *Coastal Erosion* on the different structures used to prevent wind and water erosion. Answer the following questions after you have finished watching.

1. How does beach grass help prevent wind erosion?

____

____

2. Match each word with its definition.

breakwater	a natural way to slow erosion of sand
seawall	barriers that are built in the water to break the waves
beach grass	barriers that are built on the coast to protect sand and buildings

Crosscutting Concepts

Stability and Change

3. How can people help prevent changes to Earth's surface?

____

____

Wind Erosion

Investigate how wind can affect the soil where farmers grow crops by conducting the simulation. Answer the questions after you have finished.

4. What kind of windbreaks can this farmer use?

____

____

5. The moving arrows represent wind. Why do they change in size?

____

____

6. Which windbreak protected the soil the best? Why do you think so?

______

______

Ways to Prevent Land Erosion

Explore the Digital Interactive *Ways to Prevent Land Erosion* on how people can build and plant in order to slow erosion. Answer the questions after you have finished.

7. How does planting fields in curved rows slow soil erosion?

______

______

______

8. How does a breakwater slow water erosion?

______

______

______

Science and Engineering Practices

Complete the "I can . . . " statements.

I can construct explanations ______________________

____

____

I can design solutions ______________________

____

____

Use examples
from the lesson to
explain what you
can do!

 Name ______________________ Date __________

Research, Investigate, and Communicate

Inquiry Activity

Designing a Way to Reduce Coastal Erosion

Materials

- ☐ tray
- ☐ sand
- ☐ tape
- ☐ plastic
- ☐ cloth
- ☐ grass
- ☐ twigs
- ☐ classroom materials
- ☐ safety goggles
- ☐ straw

You will design and test a way to slow down wind erosion at the beach.

Define a Problem How can you reduce wind erosion at the beach?

____

____

____

Design a Solution

1. Create a beach model at one end of the tray.
2. Think about what you have learned about wind erosion and how to slow it down.
3. Use the materials provided by your teacher to create a windbreak.

BE CAREFUL Wear safety goggles.

4. **Test** Test your windbreak by blowing on the sand through a straw. Make any changes that you would like.

5 Retest your windbreak.

Communicate Information

1. Did your windbreak reduce wind erosion? How do you know?

____

____

____

2. Draw and label your final wind erosion solution in the box below.

3. Describe how you could make your solution better.

____

____

 Name ______________________ Date ____________

Performance Task

Compare Solutions

You will compare how well your windbreak slows erosion to how well your classmate's windbreak slows erosion.

Materials

- [ ] your beach model with windbreak
- [ ] safety goggles
- [ ] hair dryer

Make a Prediction Who has created a better solution to slow or prevent wind erosion?

____

____

____

Carry Out an Investigation

BE CAREFUL Wear safety goggles.

1. Watch as your teacher uses the hair dryer to simulate wind on your beach model.

Which solution protects the beach from wind erosion the best?

2 In the box below, draw and write about the erosion that happened to your beach model.

3 Watch as your teacher uses the hair dryer to simulate wind on your classmate's beach model.

EVALUATE Name ____________________ Date __________

4 In the box below, draw and write about the erosion that happened to your classmate's beach model.

5 **Draw Conclusions** Who had the best windbreak to slow or stop wind erosion? How do you know?

____

____

____

____

Essential Question

How can people slow the changes to Earth's surface?

Look back at the photo of a sandy beach. Use what you have learned to tell why there is a fence in the middle of the sand.

____

____

____

____

____

Science and Engineering Practices

I did construct explanations.
I did design solutions.

Now that you're done with the lesson, rate how well you did.

Rate Yourself

Color in the number of stars that tell how well you did construct explanations and design solutions.

 Name ______________________ Date __________

Earth's Surface Changes

Performance Project

Reducing Flood Damage

You will design a solution for flooding and compare it to a classmate's solution.

Define a Problem How can you slow down the flooding of Maya's favorite wildflower field?

____

____

Design a Solution

1. Reread pages 16 and 17 in *Our Changing Earth* on flooding.

2. **Record Data** Draw and label a picture of your solution in the box below.

Name ______________________ Date __________

People can help slow the changes to Earth's surface!

3 Talk to your classmates about their designs. How does your solution compare to theirs?

___

___

___

___

___

___

___

___

___

___

Explore More in My World

Did you learn the answers to all of your questions from the beginning of the module? If not, how could you design an experiment to help answer them?

 Name ______________________ Date __________

Living Things in Habitats

Science in My World

Look at the photo of a rain forest. Look at all of the bright, green leaves. What questions do you have about the rain forest?

____

____

____

____

____

Key Vocabulary

Look and listen for these words as you learn about living things in habitats.

Arctic	desert	food chain
forest	grassland	habitat
ocean	pond	predator
prey	shelter	

Name ______________________ Date ____________

What living things are found in a rain forest?

RUBY
Veterinarian

Ruby wants to be a veterinarian. A veterinarian is a person who cares for the health of pets and other animals. Veterinarians need to know about an animal's habitat in order to treat the animal properly. Ruby wants to know what living things are found in a rain forest. Draw an example of a living thing from the rain forest below.

Science and Engineering Practices

I will carry out investigations.
I will plan investigations.

 Name ________________________ Date __________

Habitats

Habitats

Three friends are talking about where habitats are found. Which friend has the best idea about habitats?

Darius: *I think habitats are found on land.*

Aiko: *I think habitats are found in water.*

Nancy: *I think habitats are found on land or in water.*

Explain your thinking.

____

____

____

Science in My World

Look at this photo. You will see hills, houses, and roads. Who do you think lives in these places? What questions do you have?

______

______

______

Essential Question

What is a habitat?

Science and Engineering Practices

I will carry out investigations.

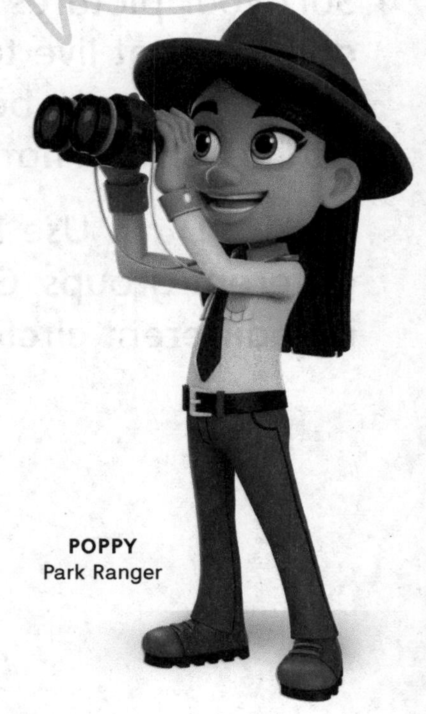

My job as a park ranger lets me learn all about plants and animals and the places where they live.
POPPY
Park Ranger

 Name ______________________ Date __________

Inquiry Activity
Living Things in Habitats

Where do plants and animals live? Look through magazines to find out.

Materials
- ☐ nature magazines
- ☐ scissors
- ☐ glue or tape

Make a Prediction What kinds of living things will you find?

____

____

Carry Out an Investigation

1. Look through magazines to find different kinds of plants and animals.
2. Cut out the pictures.
3. Sort your pictures into groups. Do you have animals that live together? Do you have plants that may be eaten by those animals? Do they live in hot or cold areas?
4. **Record Data** Use the circles to represent different groups. Glue the pictures by group into different circles.

Communicate Information

1. **Draw Conclusions** How many different groups do you have? What do those plants and animals have in common?

____

____

____

____

____

____

Name ______________________ Date ____________ EXPLAIN

Obtain and Communicate Information

Vocabulary

Use these words when explaining habitats.

habitat	predator	prey
shelter	food chain	

Plant and Animal Habitats

Read pages 14–23 in *Plant and Animal Habitats.* Answer the following questions after you have finished reading.

Fill in the blanks.

1. A habitat has space, ____________________,

   and ____________________ for the things that live there.

2. Circle the statement if it is true.
   Place an X over the statement if it is false.

   Animals use their shelter to play with their friends.

   Some animals use plants for shelter and food.

3. Can a plant that lives in a dry habitat live in a wet habitat? Explain.

______

______

Habitats and Living Things

Watch *Habitats and Living Things* on how a habitat is just right for the plants and animals that live there. Answer the questions after you have finished watching.

4. How are the two water habitats, oceans and ponds, different from each other?

______

______

______

______

5. How can plants and animals live in a cold, dry place?

______

______

______

______

Habitats

Explore the Digital Interactive *Habitats* on many different habitats found on Earth. Answer the questions after you have finished.

6. In the top row of the table, list habitats you have learned about. In the bottom row, write something you have learned about each habitat.

7. Since a desert gets very little rain, how do plants survive there?

____

____

Science and Engineering Practices

Complete the "I can . . . " statement.

I can carry out investigations ______________________

____

____

Use examples from the lesson to explain what you can do!

Research, Investigate, and Communicate

Plants and Animals Depend on Each Other

Read *Plants and Animals Depend on Each Other* on how plants and animals are connected in a food chain. Answer the questions after you have finished reading.

1. Match each word with its definition.

food chain	an animal that is hunted
prey	the order of how living things get the food they need
predator	an animal that hunts other animals

Fill in the blank.

2. Plants need sunlight in order to grow, and ______________ eat plants in order to live.

3. Where do plants get the energy to grow?

____

____

____

Inquiry Activity
Food Chains

Materials
- ☐ yellow, green, red, brown, and orange strips of paper
- ☐ crayons or colored pencils
- ☐ glue

You will create a food chain using different colored strips of paper to represent different plants and animals.

Make a Prediction How many living things will be in your food chain?

____

____

Make a Model

1. All food chains begin with the Sun. Draw the Sun on the yellow strip of paper.
2. On the green strips of paper, draw different plants.
3. On the red, brown, and orange strips, draw different animals in your food chain.
4. Lay the strips out on the table in order.
5. Glue the strips together to create a food chain.

Communicate Information

4. Tell the order of your food chain starting with the Sun.

____

____

5. What animals in your food chain are prey?

____

____

6. What animals in your food chain are predators?

____

7. **Draw Conclusions** Compare your food chain with a classmate's food chain. How are they alike? How are they different?

____

____

____

____

Crosscutting Concepts

Structure and Function

8. Why are food chains important in a habitat?

____

____

____

 Name ____________________ Date __________

Performance Task

Design a Habitat for Yourself

Materials

- [ ] crayons or colored pencils

You will use what you have learned about habitats to create one for yourself.

Make a Prediction What things do you need in your habitat to survive?

____

____

____

Make a Model

1. **Record Data** Design a habitat that you live in and label all the things you need to survive.

Name ______________________ Date ____________ EVALUATE

Essential Question

What is a habitat?

Think about the photo of the neighborhood habitat at the beginning of the lesson. Use what you have learned to describe what you see and tell how it is important for a human habitat.

____

____

____

Science and Engineering Practices

I did carry out investigations.

Now that you're done with the lesson, rate how well you did.

Rate Yourself

Color in the number of stars that tell how well you did carry out investigations.

Name ____________________ Date __________

Forests and Grasslands

Forests and Grasslands

Three friends are looking at pictures of animals in their habitats. Which friend has the best idea about the kinds of animals in forests and grasslands?

Mila: *I think forests and grasslands have the same kinds of animals.*

Violet: *I think forests and grasslands have some of the same kinds of animals.*

Sadie: *I think forests and grasslands have different kinds of animals.*

Explain your thinking.

____

____

Science in My World

Look at this habitat with a lot of trees. What do you think might live there? What questions do you have?

____

____

____

____

Essential Question

What lives in forests and grasslands?

Did you know there are different kinds of forests? Let's investigate more to learn about forests and grasslands.

Science and Engineering Practices

I will carry out investigations.

I will plan investigations.

 Name ______________________ Date __________

Inquiry Activity
Pill Bug Habitat

What is life like in a small forest?

Make a Prediction What will happen to a pill bug in its small forest habitat?

____

____

Materials

- ☐ jar
- ☐ soil
- ☐ dead leaves
- ☐ rocks
- ☐ water
- ☐ pill bug
- ☐ plastic wrap
- ☐ tape or rubber band

Make a Model

1. Place the soil and dead leaves in the bottom of the jar.
2. Arrange the rocks around the soil, and then lightly water the soil.
3. Carefully add the pill bug. Cover the jar with plastic wrap, and secure it with tape or a rubber band. Poke holes in the plastic wrap.
4. Place the jar near the window and observe.
5. **Record Data** Use the table to draw or write your observations of the pill bug.

Day	My Observations

Communicate Information

1. How did the pill bug get what it needed to survive?

____

____

____

____

 Name ______________________ Date __________

Obtain and Communicate Information

Vocabulary

Use these words when explaining forests and grasslands.

forest grassland

Types of Habitats

Watch *Types of Habitats* on forests and grasslands. Answer the questions after you have finished watching.

1. Complete the table by listing animals that live in each habitat.

rain forest	woodland forest	prairie	savanna

2. Prairies and savannas are two different types of ____________________.

Living Things in Habitats

Read *Living Things in Habitats* on forests and grasslands. Answer the questions after you have finished reading.

Fill in the blank with the name of the correct habitat.

3. This habitat is hot and dry with lots of room to roam. It is a ______________________.

4. This habitat has lots of trees that provide fruit and nuts for its animals. It is a ______________________.

5. This habitat is hot and damp with many bright green plants. It is a ______________.

Inquiry Activity

Living Things in a Forest

Materials

- ☐ pill bug habitat
- ☐ earthworm
- ☐ hand lens

You will plan and carry out an investigation to see what happens when you add an earthworm to your pill bug habitat.

Make a Prediction How will the pill bug and earthworm react in their forest habitat?

____

____

Carry Out an Investigation

1. Remove the plastic wrap from your pill bug habitat.

   Write more steps to complete your investigation.

2. ______
3. ______
4. ______

Communicate Information

6. **Record Data** In the box below, draw or write your observations of the earthworm and the pill bug.

Name ______________________ Date ____________ EXPLAIN

Crosscutting Concepts

Structure and Function

7. Why do you think earthworms would be an important animal in a forest?

____

____

____

____

____

Use examples from the lesson to explain what you can do!

Science and Engineering Practices

Complete the "I can . . ." statements.

I can carry out investigations ______________________

____

I can plan investigations ______________________

____

____

 Name ______________ Date ________

Research, Investigate, and Communicate

Rainfall in the Forest

Read *Forest Habitats* on the differences between a woodland forest and a rain forest. Answer the questions after you have finished reading.

1. Name one way a woodland forest is different than a rain forest.

____

____

____

____

2. Use the rainfall amounts below to create a graph on the next page.

Average Yearly Rainfall	Amazon Rain Forest	White River National Forest
January	190 mm	50 mm
April	200 mm	40 mm
July	50 mm	70 mm
October	120 mm	50 mm

3. In what month does the White River National Forest receive the most rainfall?

____

____

4. Which habitat receives more rainfall in April? How much more?

____

____

5. Looking at just the four months on your graph, does the Amazon Rain Forest or the White River National Forest receive more rain each year? How much more?

____

____

____

____

Name _______________ Date _______________ EVALUATE

Animal Research Partner Activity

Research Look through materials provided by your teacher to learn about an animal from a forest or grassland habitat.

Ask a Question What would you like to learn about the animal?

____

____

1. **Record Data** Which animal will you write about? Describe the animal.

____

____

2. What does this animal need to survive in its habitat?

____

____

3. Write at least two other facts you would like to share about this animal.

____

____

Communicate Information

Communicate Write a story about living as the forest or grassland animal you researched.

Essential Question

What lives in forests and grasslands?

Think about the photo of a forest at the beginning of the lesson. Use what you have learned about forests and grasslands to write about the plants and animals that live there.

____

____

____

____

Science and Engineering Practices

I did carry out investigations.

I did plan investigations.

Rate Yourself

Color in the number of stars that tell how well you did carry out and plan investigations.

Now that you're done with the lesson, rate how well you did.

 Name ______________________ Date __________

Water Habitats

Water Habitats

Some organisms live in water.
Circle the things they need to live.

Food	The right temperature
Air	Sunlight
Noise	A place to raise young
Shelter	People

Explain your thinking.

____

____

____

____

Name ______________________ Date __________

Science in My World

Look at the photo of the water habitat. What do you see in this photo? What questions do you have?

____

____

____

____

____

Essential Question

What lives in water habitats?

Science and Engineering Practices

I will carry out investigations.

As a park ranger, I can tell you that some habitats in our parks are under water!

Name ______________________ Date __________

Inquiry Activity
Brine Shrimp

Materials

- ☐ 2 containers
- ☐ water
- ☐ measuring cup
- ☐ salt
- ☐ measuring spoons
- ☐ spoon
- ☐ brine shrimp eggs
- ☐ hand lens
- ☐ plastic wrap
- ☐ rubber band

What kind of habitat does a brine shrimp need to survive?

Make a Prediction Do brine shrimp eggs need salt water or fresh water to hatch?

____

____

Carry Out an Investigation

1. Fill each container with two cups of water. Add two teaspoons of salt to one container, and stir with a spoon.
2. Add ¼ teaspoon of brine shrimp eggs to each container.
3. Using the hand lens, observe the eggs every day for one week.

Name ______________________ Date ____________ EXPLORE

Communicate Information

1. **Record Data** Record your observations in the table. When did the eggs hatch?

	Salt Water	Fresh Water
Day 1		
Day 2		
Day 3		
Day 4		

Draw Conclusions

2. Which water habitat do the brine shrimp prefer?

____

____

3. How do your containers of brine shrimp compare to those of your classmates?

____

____

 Name ______________________ Date __________

Obtain and Communicate Information

Vocabulary

Use these words when explaining water habitats.

ocean pond

Oceans and Ponds

Watch *Oceans and Ponds* on the differences between these two water habitats. Answer the questions after you have finished watching.

1. What is a pond?

____

____

2. What kinds of plants and animals live in ponds?

____

____

____

____

Name ______________________ Date ____________ EXPLAIN

Living Things in Oceans and Ponds

Read *Living Things in Oceans and Ponds* on the differences between the plants and animals living in these two water habitats. Answer the questions after you have finished reading.

Fill in the blanks.

3. Seagrass and kelp are two ocean plants

that provide ____________________ and

____________________ for many animals living in the ocean.

4. Draw a picture of a pond or ocean habitat. Label the plants and animals you have included in your drawing.

 Name ________________ Date ________

Inquiry Activity
Floating Fish

You will investigate part of a fish's body that helps it survive in its habitat.

Materials

- ☐ tray
- ☐ water
- ☐ water bottle

Make a Prediction What makes a fish able to float?

____

____

Carry Out an Investigation

1. Fill the tray with water.
2. Fill the water bottle halfway with water, screw the cap back on, and place the bottle in the tray. Observe what happens.
3. Pour most, but not all, of the water out of the bottle. Place the bottle in the tray again. Observe any changes.

Communicate Information

5. **Record Data** Record your observations in the table. Draw what you observed.

Bottle with Equal Amounts of Air and Water	Bottle with Little Water and Lots of Air

6. **Draw Conclusions** What happened to the bottle when you poured out most of the water?

____

____

____

Crosscutting Concepts

Structure and Function

7. How does the swim bladder help a fish survive in the water?

____

____

____

Coral Reefs

Explore the Digital Interactive *Coral Reefs* on another water habitat. Answer the questions after you have finished.

8. What are some plants and animals that can be found in a coral reef habitat?

____

____

____

9. Is coral a plant or an animal?

____

____

____

Use examples from the lesson to explain what you can do!

Science and Engineering Practices

Complete the "I can . . . " statement.

I can carry out investigations ______________________________

____

____

Research, Investigate, and Communicate

River and Stream Habitats

Read *River Habitats* on a water habitat that is always moving. Answer the questions after you have finished reading.

1. What animals live in rivers?

____

____

____

____

2. How are rivers different from lakes and ponds?

____

____

____

____

____

Performance Task

Materials
- ☐ shoe box
- ☐ variety of art supplies

Animal Research Partner Activity

Research Working with a partner, you will research a water habitat and create a model.

Ask a Question What question will your research help to answer?

____

____

1. **Record Data** Which water habitat will you make a model of?

____

2. What two animals will you include in your model?

____

3. What two plants will you include in your model?

____

4. Use a shoe box and a variety of art supplies to create a model of your water habitat.

Name ____________________ Date __________ EVALUATE

Essential Question

What lives in water habitats?

Think about the photo of a coral reef at the beginning of the lesson. Use what you have learned to tell about water habitats.

____

____

____

____

____

Science and Engineering Practices

I did carry out investigations.

Rate Yourself

Color in the number of stars that tell how well you did carry out investigations.

Now that you're done with the lesson, rate how well you did.

 Name ____________ Date ______

Hot and Cold Deserts

Deserts

Three friends are reading a book about deserts. Which friend has the best idea about deserts?

Mason: *Deserts are found in warm areas.*

Henry: *Deserts are found in cold areas.*

Dominic: *Deserts are found in warm areas and in very cold areas.*

Explain your thinking.

____

____

____

Name ______________________ Date ____________ ENGAGE

Science in My World

Look at the photo of the mountains and snow. What kind of habitat do you think this is? What questions do you have?

____

____

____

____

Essential Question

What can live in hot and cold deserts?

Science and Engineering Practices

I will plan and carry out investigations.

Brrrrr, it sure looks cold with all of that ice and snow. What kinds of plants and animals can live in a habitat like this?

 Name ______________________ Date __________

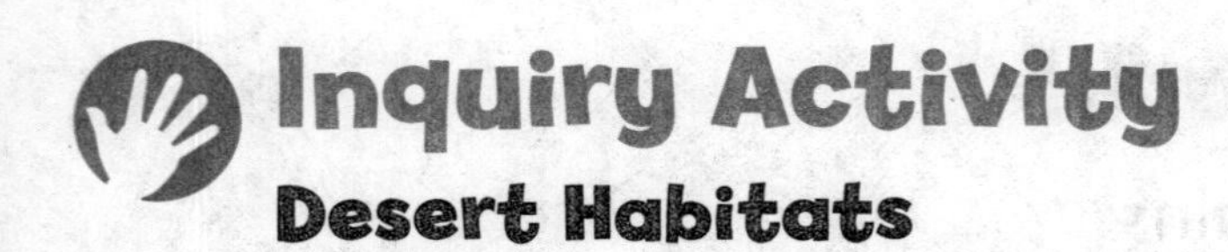

Inquiry Activity
Desert Habitats

How do animals survive in very hot or very cold habitats?

Make a Prediction How are animals able to live in cold deserts?

____

____

Carry Out an Investigation

Investigate animals in hot and cold habitats by conducting the simulation. Answer the questions after you have finished.

Communicate Information

1. **Record Data** Record your observations of animals in a hot desert.

	Scorpion	Fox	Owl
Day			
Night			

2. Now observe how animals act during the summer and winter in a cold desert. Record your observations in the table below.

	Lemming	Bumblebee	Caribou
Summer			
Winter			

3. **Draw Conclusions** What do most of the animals in a hot desert do during the middle of the day? Why?

____

____

4. What do most animals in a cold desert do in the winter? Why?

____

____

 Name ______________________ Date __________

Obtain and Communicate Information

Vocabulary

Use these words when explaining hot and cold deserts.

desert Arctic

Extreme Habitats

Read pages 14–23 in *Extreme Habitats* on how plants and animals are able to live in very hot and very cold conditions. Answer the questions after you have finished.

1. Circle the statement if it is true. Place an X over the statement if it is false.

   The Arctic is a very cold desert in the northern part of Earth.

   Plants in cold deserts grow very quickly.

   A desert is very dry land that does not have many plants.

2. What is one way Arctic animals stay warm?

____

____

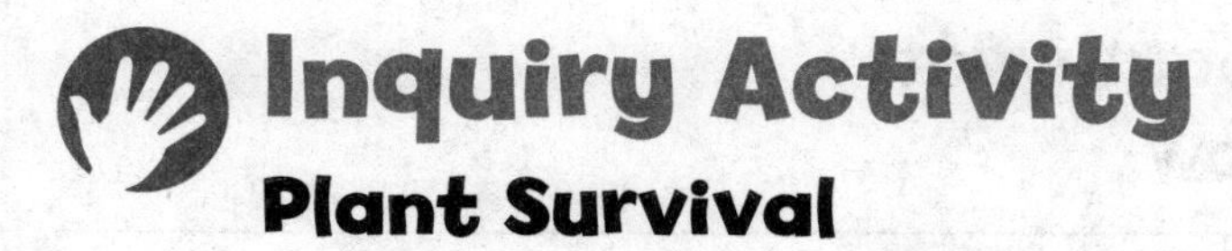

Inquiry Activity
Plant Survival

Materials
- ☐ cactus
- ☐ lily

You will plan and carry out an investigation about plants in a desert habitat.

Make a Prediction What will happen to a cactus and a lily in a desert habitat where there is very little water?

____

____

Carry Out an Investigation

1. Write the steps of your investigation on the lines below.

____

____

____

____

____

____

____

 Name ______________________ Date __________

3. Record Data Draw or write about what you observed. Use the box below.

Plants Before the Investigation	Plants After the Investigation

4. Draw Conclusions How did the cactus look at the end of your investigation?

___

___

5. How did the lily look at the end of your investigation?

___

___

Crosscutting Concepts

Structure and Function

6. How was the cactus able to survive without water?

___

___

___

Extreme Habitats

Watch *Extreme Habitats* on how plants and animals live in hot and cold deserts. Answer the questions after you have finished watching.

Fill in the blanks.

7. A polar bear is able to stay warm in the Arctic because of its

   ________________.

8. Animals that live in deserts can get

   water by ________________.

Use examples from the lesson to explain what you can do!

Science and Engineering Practices

Complete the "I can . . . " statement.

I can plan and carry out investigations ________________

____

____

____

 Name ______________________ Date __________

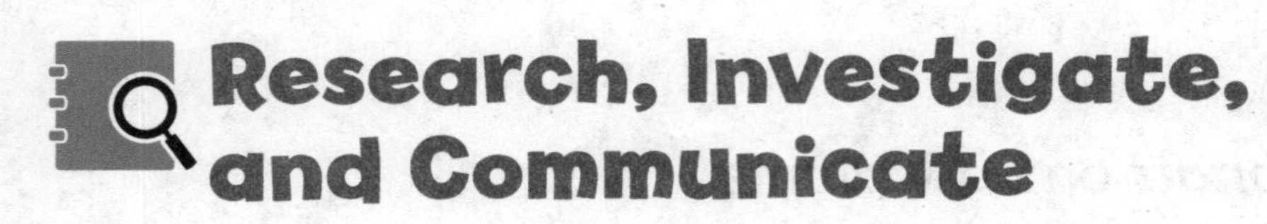

Research, Investigate, and Communicate

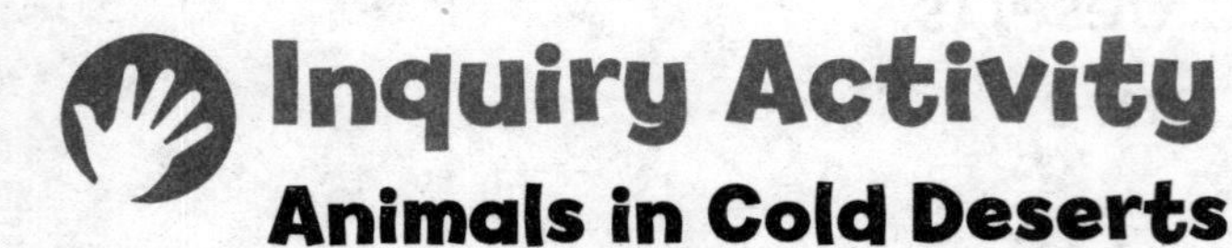

Inquiry Activity

Animals in Cold Deserts

Materials

- [ ] hot water
- [ ] foam cup
- [ ] regular cup
- [ ] 2 thermometers

You will observe how animals stay warm in cold deserts.

Make a Prediction Will hot water stay warm longer in a foam cup or a regular cup?

____

____

Carry Out an Investigation

1. Watch as your teacher pours hot water into a foam cup and into a regular cup.
2. **Record Data** Measure the temperature in each cup. Record your data in the table.

Name ______________________ Date __________ ELABORATE

	Regular Cup	Insulated Cup
Beginning temperature		
5 minutes		
10 minutes		
15 minutes		
20 minutes		

Communicate Information

1. **Communicate** Which container kept the water warm longer? Why?

____

____

Crosscutting Concepts

Structure and Function

2. How do you think Arctic animals stay warm in their cold habitat?

____

____

____

 Name ______________________ Date __________

Performance Task
Habitat Wrap-Up

Materials
- ☐ construction paper
- ☐ crayons or colored pencils

You will show what you have learned about hot and cold deserts.

Make a Prediction How can you make a model to show what you know about hot and cold deserts?

____

____

____

Make a Model

1. Use the paper provided by your teacher. Fold the paper in half and then open it back up. Draw a picture of a hot desert on the left side and a cold desert on the right side.
2. Turn the paper over and record facts about plants, animals, and the weather conditions of a hot and cold desert.

Name ______________________ Date ____________ EVALUATE

Essential Question

What can live in hot and cold deserts?

Think about the photo of the Arctic desert at the beginning of the lesson. Use what you have learned to discuss plants and animals that live in a cold desert.

____

____

____

____

Science and Engineering Practices

I did plan and carry out investigations.

Rate Yourself

Color in the number of stars that tell how well you did plan and carry out investigations.

Now that you're done with the lesson, rate how well you did.

 Name ______________________ Date __________

Living Things in Habitats

Performance Project

Designer Habitat

You will design a habitat for a mystery animal.

Define a Problem What kind of habitat does a mystery animal need?

____

Design a Solution

A mystery animal needs a place to live. The animal is big and has yellow fur with black spots. It hunts monkeys for food. It swims well, but it can also walk long distances.

Communicate Information

1. What is the mystery animal?

____

2. Where does the mystery animal live?

____

(1) On the next page, design a habitat for this mystery animal. Be sure to include the animal and what it needs to survive in its habitat.

Name ______________________ Date ____________

Living things find what they need to survive in their habitat.

Explore More in My World

Did you learn the answers to all of your questions from the beginning of the module? If not, how could you design an experiment to help answer them?

 Name ______________________ Date __________

Plants and Their Needs

Science in My World

Look at the photo of the flowers and plants. How many different plants do you see? What questions do you have about the garden?

____

____

____

____

____

Key Vocabulary

Look and listen for these words as you learn about plants and their needs.

dispersal	flower	fruit
germinate	leaves	nutrient
pollen	pollination	root
seed	seedling	stem

Name ______________________ Date ____________

What do plants need in order to live and grow?

KAYLA
Landscape Architect

Kayla wants to be a landscape architect. Landscape architects design the placement of plants outside buildings and houses to make them look nice. They also decide the best plants to use in the area they are designing. Kayla wants to know what plants need to live and grow. Write or draw what you think.

Science and Engineering Practices

I will plan and carry out investigations.
I will develop and use models.

 Name ______________________ Date __________

Plants Need Water

Do All Plants Need Water?

Two friends are talking about the needs of plants. Which friend has the best idea about plants and water?

Isabella: *I think all plants need water.*

Faye: *I think most plants need water. There are some plants that do not need water.*

Explain your thinking.

____

____

Name ______________________ Date ____________

Science in My World

Look at the photo of the garden. What do you notice about what is growing there? What questions do you have?

____

____

____

____

____

Essential Question

Why do plants need water?

Science and Engineering Practices

I will plan and carry out investigations.

Healthy plants are important to my insect friends. Let's investigate why plants need water.
OWEN
Entomologist

Name ______________________ Date __________

Inquiry Activity

Seeds Needs

Can a seed begin to grow in a plastic bag?

Make a Prediction Will a seed begin to grow in a plastic bag with or without water?

____

____

Materials

- ☐ 2 paper towels
- ☐ tablespoon
- ☐ water
- ☐ 2 plastic bags
- ☐ 6 bean seeds
- ☐ hand lens
- ☐ permanent marker

Carry Out an Investigation

1. Fold each paper towel into quarters. Put two tablespoons of water onto one of the folded towels. Place that wet towel into a plastic bag, and label the bag "Water."
2. Place the dry paper towel in the other bag, and label it "No Water."
3. Place three seeds into each bag, seal the bags, and place them in a sunny spot.
4. **Record Data** Observe the seeds every day for 5 days. Record your observations by drawing pictures and labeling them in the table.

	Day 1	Day 2	Day 3	Day 4	Day 5
Water					
No Water					

Communicate Information

1. Did any of the seeds change? How?

____

____

____

2. Why do you think the seeds changed?

____

____

____

Name ______________________ Date __________

Obtain and Communicate Information

Vocabulary

Use these words when explaining why plants need water.

flower	stem	leaves
root	germinate	

Parts of a Plant

Explore the Digital Interactive *The Parts of a Plant* on how plant parts help a plant live and grow. Answer the questions after you have finished.

1. Match each plant part with its function.

stem	take in air and use sunlight to make food
leaves	take in water and minerals
flower	allows food to travel through the plant
roots	makes seeds and fruit

2. Which two parts help move water through a plant?

Inquiry Activity

Do Plants Need Water to Grow?

Materials

- ☐ two potted plants
- ☐ tape for labeling the plants
- ☐ permanent marker
- ☐ water

You will plan and carry out an investigation to see if plants need water to continue to grow.

Make a Prediction Will a plant continue to grow if it is not watered?

____

____

Carry Out an Investigation

Plan the investigation by putting the following steps in the correct order.

☐ Water the plant labeled "Water", as needed, for one week. Do not water the plant labeled "No Water" for one week.

☐ Draw a picture in the table at the top of the next page of how your plants look at the beginning of the investigation.

☐ Label one plant with the word "Water" and the other with the words "No Water."

☐ Get two plants from your teacher.

Name ______________________ Date __________

"Water"	"No Water"

Draw Conclusions

3. What happened when you carried out your investigation? Draw what your plants looked like after one week.

"Water"	"No Water"

Name ______________________ Date ____________ EXPLAIN

Crosscutting Concepts

Cause and Effect

4. What happens when plants do not get water over a period of time?

____

____

____

Watch *Roots Take In Water.* Answer the following questions after you have finished watching.

5. Name two reasons that roots grow down into soil.

____

____

____

6. One job of plant roots is to bring water into a plant. Name another job of plant roots.

____

____

____

Name ______________________ Date ____________

7. Circle the statement that is true.
Place an X over the statement that is false.

Flowers grow upwards toward rain clouds.

Flowers grow upwards toward the sun.

Science and Engineering Practices

Complete the "I can . . . " statement.

I can plan and carry out investigations ____________________

____

____

____

____

Use examples from the lesson to explain what you can do!

Research, Investigate, and Communicate

Seeds Need Water

Read *Seeds Need Water* on what seeds need to grow into plants. Answer the questions after you have finished reading.

1. What does a seed need to germinate?

____

____

2. What are the first plant parts you see after a seed germinates?

____

____

3. How does a seed germinate in hydroponics?

____

____

____

____

____

EVALUATE Name ______________________ Date __________

Performance Task

Draw a Picture of a Flowering Plant

Materials

- [ ] crayons or colored pencils

You will draw a before-and-after picture of a flowering plant that has not been watered.

Make a Model

1. Draw and label a flowering plant that has been watered.

2. Draw and label a flowering plant that has not been watered.

Essential Question

Why do plants need water?

Think about the photo of the garden at the beginning of the lesson. Use what you have learned to explain why plants need water.

____

____

____

____

____

Science and Engineering Practices

I did plan and carry out investigations.

Rate Yourself

Color in the number of stars that tell how well you did plan and carry out investigations.

Now that you're done with the lesson, rate how well you did.

Name ______________________ Date __________

Plants Need Light

Do All Plants Need Light?

Two friends are talking about the needs of plants. Which friend has the best idea about plants and light?

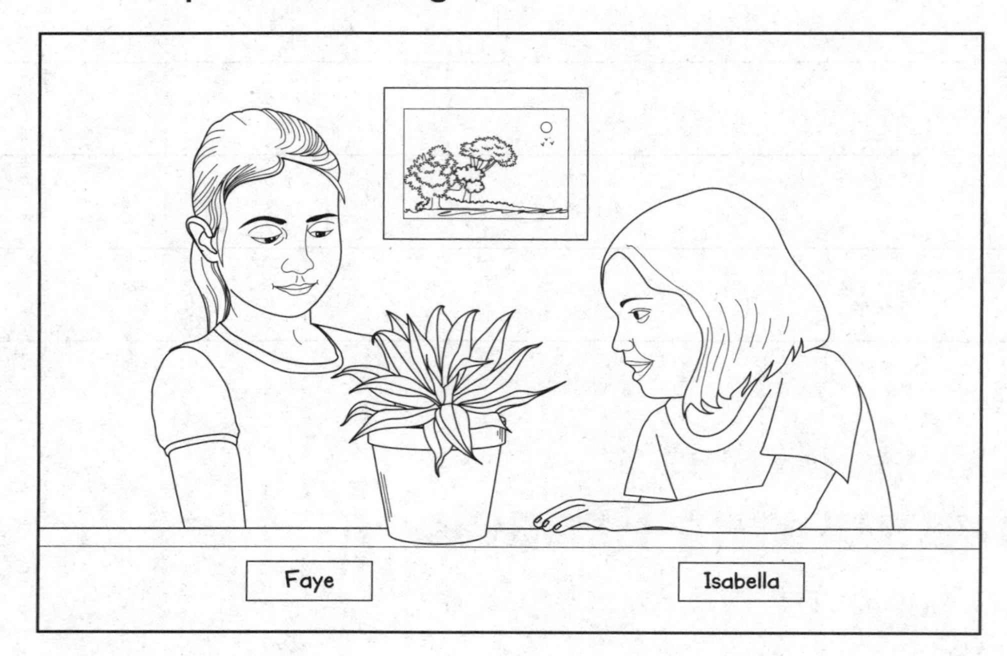

Isabella: *I think all plants need light.*

Faye: *I think most plants need light. There are some plants that live only in the dark.*

Explain your thinking.

____

____

____

____

Science in My World

Look at the photo of plants growing in the ball. What do you notice about the container they are growing in? What questions do you have?

____

____

____

____

Essential Question

Why do plants need light?

I am curious what else plants need to stay healthy. Do plants need sunlight?

Science and Engineering Practices

I will plan and carry out investigations.

Name ______________________ Date __________

Inquiry Activity
Plants and Sunlight

Materials
- ☐ 2 potted plants
- ☐ aluminum foil
- ☐ tape
- ☐ hand lens

What will happen to a plant after one week with no sunlight on its leaves?

Make a Prediction Do a plant's leaves need sunlight for the plant to grow?

____

____

____

Carry Out an Investigation

1. Tape foil over all the leaves of one of the potted plants.
2. Place both plants in a sunny location, and water them when needed.
3. After a week, remove the aluminum foil from the leaves.
4. Using a hand lens, observe the differences between the leaves on the two plants.

Name ______________________ Date ____________ EXPLORE

Communicate Information

1. **Record Data** Draw a picture of the leaves that were covered in foil. Then draw a picture of the leaves that were not covered in foil.

Leaves That Were Covered	Leaves That Were Not Covered

2. How do your drawings compare to your classmates' drawings?

____

____

3. Look at your prediction. Were you able to discover if a plant's leaves need sunlight to grow?

____

____

 Name ______________________ Date __________

Obtain and Communicate Information

Vocabulary

Use these words when explaining why plants need sunlight.

nutrient mineral

How Plants Use Their Parts to Live and Grow

Read pages 14–23 in *How Plants Use Their Parts to Live and Grow.* Answer the questions after you have finished.

1. What do plants use to make their own food?

____

____

2. Food provides all living things with

_______________ to live and grow.

Name ________________________ Date ____________ EXPLAIN

Inquiry Activity

How Leaves Help a Plant Get Light

Materials

- ☐ pictures of leaves

You will observe how the shape and size of a leaf affect how much light a plant can collect.

Make a Prediction What shape and size of a leaf help a plant collect the most sunlight?

____

____

Carry Out an Investigation

1. Look at the pictures, and compare the leaves of different plants.
2. Hold out your hand in a sunny spot. Open and close your hand, then turn it from side to side.
3. Compare the shape of a leaf to the shape of your hand when you hold it flat.

Communicate Information

3. **Record Data** Draw a picture of your hand when it gets the most sunlight. Draw a picture of a leaf that gets the most sunlight.

Hand Drawing	Leaf Drawing

Crosscutting Concepts

Structure and Function

4. What shape and size of a leaf get the most sunlight for its plant?

____

____

____

How Do Plants Make Food?

Explore the Digital Interactive *How Do Plants Make Food?* on the different parts of a plant. Answer the questions after you have finished.

Choose the best word to complete each statement.

stem leaves food air

5. ______________ change air, water, and light into food for plants.

6. Plants take in carbon dioxide and release oxygen back into the

______________.

7. Plants can pass stored ______________ on to other living things.

8. Food produced by leaves moves to other parts of the plant through the

______________.

 Name ______ Date ______

Crosscutting Concepts
Structure and Function

9. Why is a leaf called the "food factory" of a plant?

______
______
______

Science and Engineering Practices

Complete the "I can . . . " statement.

I can plan and carry out investigations ______
______
______

Use examples from the lesson to explain what you can do!

Research, Investigate, and Communicate

Plant Care Research

Research You will research how to care for a plant of your choice.

Ask a Question What question will your research help to answer?

______

______

Research how to care for the plant. Fill in the boxes below with information about the plant.

Needs	Habitat	Important Facts

Communicate Information

1. Use your research to write a paragraph explaining how to care for your plant.

2. What does this plant need that all types of plants need?

Performance Task

What Do You Still Wonder About Plants?

You will list questions you still have about plants and plan an investigation that will help you get answers to your questions.

Ask a Question What else do you want to know about plants?

____

____

Carry Out an Investigation

1. Work with a partner to make a list of questions you still have about plants' need for water and sunlight.

____

____

____

2. Choose one question you wrote. Make a list of possible ways you could answer that question.

____

____

____

3 Choose one idea from your list. Write the steps you would take to complete this investigation.

Essential Question

Why do plants need light?

Think about the photo of a terrarium you saw at the beginning of the lesson. Use what you have learned to explain why plants need light.

____

____

____

____

____

Science and Engineering Practices

I did plan and carry out investigations.

Rate Yourself

Color in the number of stars that tell how well you did plan and carry out investigations.

Now that you're done with the lesson, rate how well you did.

Name ______________________ Date __________

Plants Make More Plants

Making More Plants

Two friends are talking about what plants need to make more plants. Which friend has the best idea?

Talya: *I think some plants need animals to help them make more plants.*

Ben: *I disagree! Animals don't help plants make more plants. They eat plants!*

Explain your thinking.

____

____

____

____

Name ______________________ Date __________

Science in My World

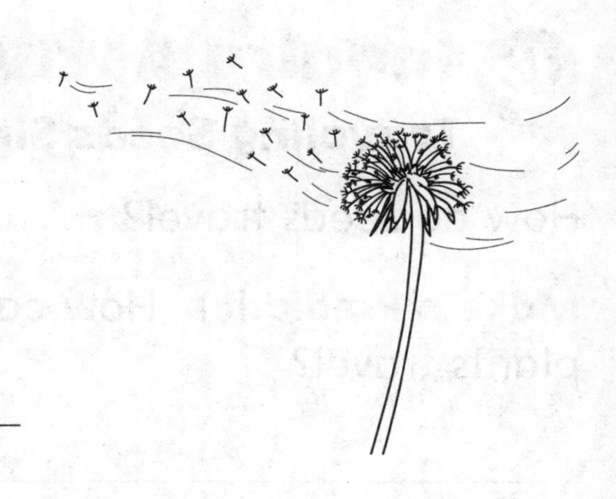

Watch the video of the dandelion blowing. What do you notice about parts of the flower? What questions do you have?

____

____

____

____

Essential Question

How do plants get help making new plants?

As an entomologist, I like investigating how bees and other insects can help plants make more plants.

Science and Engineering Practices

I will carry out investigations.
I will develop and use models.

 Name ______________________ Date __________

Inquiry Activity

Traveling Seeds Simulation

How do seeds travel?

Make a Prediction How can seeds of different plants travel?

____

____

Carry Out an Investigation

Investigate the ways seeds travel by conducting the simulation.

1. Observe how many of each type of plant grows when you first open the simulation.
2. When given the option, choose either *wind* or *animal* to see how seeds travel differently with the help of wind or an animal.
3. Observe how many of each type of plant grows with help from the wind or help from an animal.

Name ________________________________ Date ______________ EXPLORE

Communicate Information

Record Data Draw how the wind helped seeds travel. Then draw how an animal helped seeds travel.

Wind	Animal

 Name ______________________ Date __________

Obtain and Communicate Information

Vocabulary

Use these words when explaining how plants make more plants.

seed	seedling	fruit
pollen	pollination	dispersal

Making New Plants

Read pages 14–23 in *Making New Plants.* Answer the questions after you have finished.

1. Match each word with its definition.

seed	the plant part that makes seeds
seedling	the plant part that holds seeds
fruit	the plant part that grows a new plant
flower	a small, young plant

2. Why must seeds travel instead of growing right where they are?

____

____

3. Name two insects that spread pollen so plants can make more plants.

____

Seeds Move from Place to Place

Explore the Digital Interactive *Seeds Move from Place to Place* on the different ways seeds can travel. Answer the questions after you have finished.

4. Which part of the plant can grow into another plant?

____

5. Which of the following does **not** help seeds move? Circle the answer.

wind animals soil

Fill in the blank.

6. Humans can help with seed ______ by picking up a seed and planting it in a new place.

Life Cycle of a Plant

Watch *Life Cycle of a Plant* on how flowers make seeds that become plants. Answer the questions after you have finished reading.

7. Circle the statement that is true. Place an X over the statement that is false.

   A flower stores food for a plant.

   Fruit helps keep seeds safe.

8. What part do flowers play in making new plants?

   ______

   ______

9. Most fruits have seeds on the inside, but ______ are an example of a fruit with seeds on the outside.

Name ______________________ Date __________ EXPLAIN

Pollination

Read *Pollination* on the important partnership between flowers and animals. Answer the questions after you have finished reading.

10. How does pollination happen?

____

____

____

____

____

____

See how helpful my insect friends can be!

11. Why do bees fly from flower to flower?

____

____

____

____

Name ______________________ Date __________

Inquiry Activity
Insect Pollination

You will investigate how an insect pollinates a flower.

Materials

- ☐ colored construction paper
- ☐ variety of art supplies, including chenille stems and glue
- ☐ paper bag flower from your teacher

Make a Prediction What will happen to pollen on your hand when you touch flower after flower?

____

____

____

Carry Out an Investigation

1. Use construction paper and other art supplies to create a flower. Make sure your flower is colorful enough to attract all kinds of birds, bees, and butterflies.
2. Pretend that your fingers are insects.
3. Reach into the flower bag your teacher gave to you. Feed on the "flower" as an insect would.

4. When you have finished, take your hand out of the "flower" and travel to a classmate's construction paper flower to feed on it.

5. Feed on as many classmates' "flowers" as time will allow.

6. Wash your hands.

12. **Record Data** Draw a picture of your flower. Include labels that explain what happened to your flower as your classmates came to feed on it.

Communicate Information

13. What did the powdered sugar represent in this investigation?

 Name ______________________ Date __________

14. What did the candy at the bottom of the bag represent?

____

15. Was your flower pollinated? Tell how you know.

____

____

____

____

Science and Engineering Practices

Complete the "I can . . . " statements.

I can carry out investigations ______________________

____

I can develop and use models ______________________

____

Use examples from the lesson to explain what you can do!

Name ______________________ Date ____________ ELABORATE

Research, Investigate, and Communicate

Inquiry Activity

Traveling Seeds Simulation

You will revisit the simulation to observe the two different seeds up close.

Make a Prediction What shapes will you see on the seeds that travel?

____

____

____

Carry Out an Investigation

Investigate the ways seeds travel by conducting the simulation.

1. Zoom in on each of the seed types in the simulation.
2. Record your observations in the boxes. Draw each kind of seed.

Name ______________________ Date __________

Dandelion Seed

Purple Flower Seed

Crosscutting Concepts

Structure and Function

How does the shape and structure of the purple flower seed allow it to be carried by the rabbit?

____

____

Performance Task

Make a Model of a Seed

You will make a model of a seed and describe how it can be dispersed by an animal.

Make a Prediction What materials will be best to create a seed that can be dispersed by an animal?

____

Materials

- ☐ variety of art supplies including chenille stems modeling clay, paper, and glue
- ☐ large pieces of faux fur, burlap, and felt

Make a Model

1. Plan your seed by thinking about the shape of other seeds that are dispersed by animals.
2. Create your seed from the art supplies provided by your teacher.
3. Add any materials that will help your seed move.
4. **Test** Use the pieces of fabric to see if your seed would stick to an animal. Wave, shake, and twist the fabric as you walk across the classroom.

Communicate Information

1. **Draw Conclusions** Does your model seed stick to the fabric long enough to be dispersed?

____

____

 Name ____________________ Date __________

2. What could you add to your seed to make it better?

3. Draw and label your model.

4. Describe how your seed is able to be dispersed by an animal.

Name ______________________ Date ____________ EVALUATE

Essential Question

How do plants get help making new plants?

Think about the dandelion video at the beginning of the lesson. Use what you have learned to tell how plants get help making new plants.

____

____

____

____

Science and Engineering Practices

I did carry out investigations.

I did develop and use models.

Now that you're done with the lesson, rate how well you did.

Rate Yourself

Color in the number of stars that tell how well you did carry out investigations and develop and use models.

 Name ______________________ Date __________

Plants and Their Needs

Performance Project

Plants in Your Area

You will plan and carry out an investigation to see which plants grow best in your area.

Materials

- [ ] 3 different types of plants

Make a Prediction Which plants will grow best in your area?

____

____

Carry Out an Investigation

Plan your investigation by writing the steps you will follow.

____

____

____

____

____

____

Name ______________________ Date __________

Plants depend on water, light, and animals to grow!

Glue your graph here.

Record Data Use a separate piece of paper to record data about your plants' growth.

Communicate Information

1. Which plant grew the best in your area?

____

____

2. How do you know which plant grew the best?

____

____

____

____

____

Explore More in My World

Did you learn the answers to all of your questions from the beginning of the module? If not, how could you design an experiment to help answer them?

An Interview with

Dinah Zike Explaining Visual Kinesthetic Vocabulary®, or VKVs®

What are VKVs and who needs them?

"VKVs are flashcards that animate words by kinesthetically focusing on their structure, use, and meaning. VKVs are beneficial not only to students learning the specialized vocabulary of a content area, but also to students learning the vocabulary of a second language."

Dinah Zike | Educational Consultant
Dinah-Might Activities, Inc. – San Antonio, Texas

Why did you invent VKVs?

"Twenty years ago, I began designing flashcards that would accomplish the same thing with academic vocabulary and cognates that Foldables® do with general information, concepts, and ideas—make them a visual, kinesthetic, and memorable experience."

I had three goals in mind:

- **Making two-dimensional flashcards three-dimensional**
- **Designing flashcards that allow one or more parts of a word or phrase to be manipulated and changed to form numerous terms based upon a commonality**
- **Using one sheet or strip of paper to make purposefully shaped flashcards that were neither glued nor stapled, but could be folded to the same height, making them easy to stack and store**

Why are VKVs important in today's classroom?

"At the beginning of this century, research and reports indicated the importance of vocabulary to overall academic achievement. This research resulted in a more comprehensive teaching of academic vocabulary and a focus on the use of cognates to help students learn a second language. Teachers know the importance of using a variety of strategies to teach vocabulary to a diverse population of students. VKVs function as one of those strategies."

An Interview with

Dinah Zike Explaining Visual Kinesthetic Vocabulary®, or VKVs®

How are VKVs used to teach content vocabulary?

“As an example, let's look at content terms based upon the combining form *–vore*. Within a unit of study, students might use a VKV to kinesthetically and visually interact with the terms *herbivore*, *carnivore*, and *omnivore*. Students note that *–vore* is common to all three words and it means “one that eats” meat, plants, or both depending on the root word that precedes it on the VKV. When the term *insectivore* is introduced in a classroom discussion, students have a foundation for understanding the term based upon their VKV experiences. And hopefully, if students encounter the term *frugivore* at some point in their future, they will still relate the *–vore* to diet, and possibly use the context of the word's use to determine it relates to a diet of fruit.”

What organization and usage hints would you give teachers using VKVs?

“Cut off the flap of a 6” x 9” envelope and slightly widen the envelope's opening by cutting away a shallow V or half circle on one side only. Glue the non-cut side of the envelope into the front or back of student workbooks or journals. VKVs can be stored in the pocket.

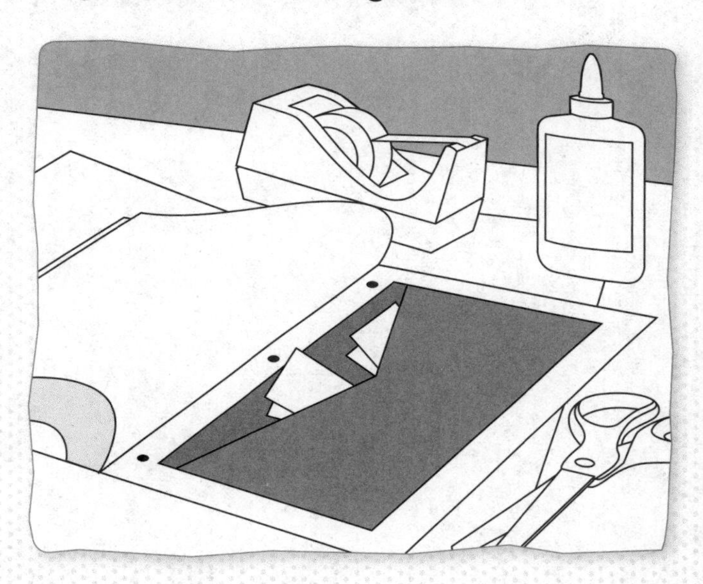

Encourage students to individualize their flashcards by writing notes, sketching diagrams, recording examples, forming plurals (radius: radii or radiuses), and noting when the math terms presented are homophones (sine/sign) or contain root words or combining forms (kilo-, milli-, tri-).

As students make and use the flashcards included in this text, they will learn how to design their own VKVs. Provide time for students to design, create, and share their flashcards with classmates.”

Dinah Zike's book Foldables, Notebook Foldables, & VKVs for Spelling and Vocabulary 4th-12th won a Teachers' Choice Award in 2011 for “instructional value, ease of use, quality, and innovation”; it has become a popular methods resource for teaching and learning vocabulary.

cut on all dashed lines

fold on all solid lines

______________ is anything that takes up space and has mass.

matter can be a gas

1. A **solid** is a state of matter has a shape of its own.
2. A **liquid** is matter that takes the shape of the container it is in.
3. A **gas** is a state of matter that does not have its own shape.

Dinah Zike's Visual Kinesthetic Vocabulary®

Properties of Matter

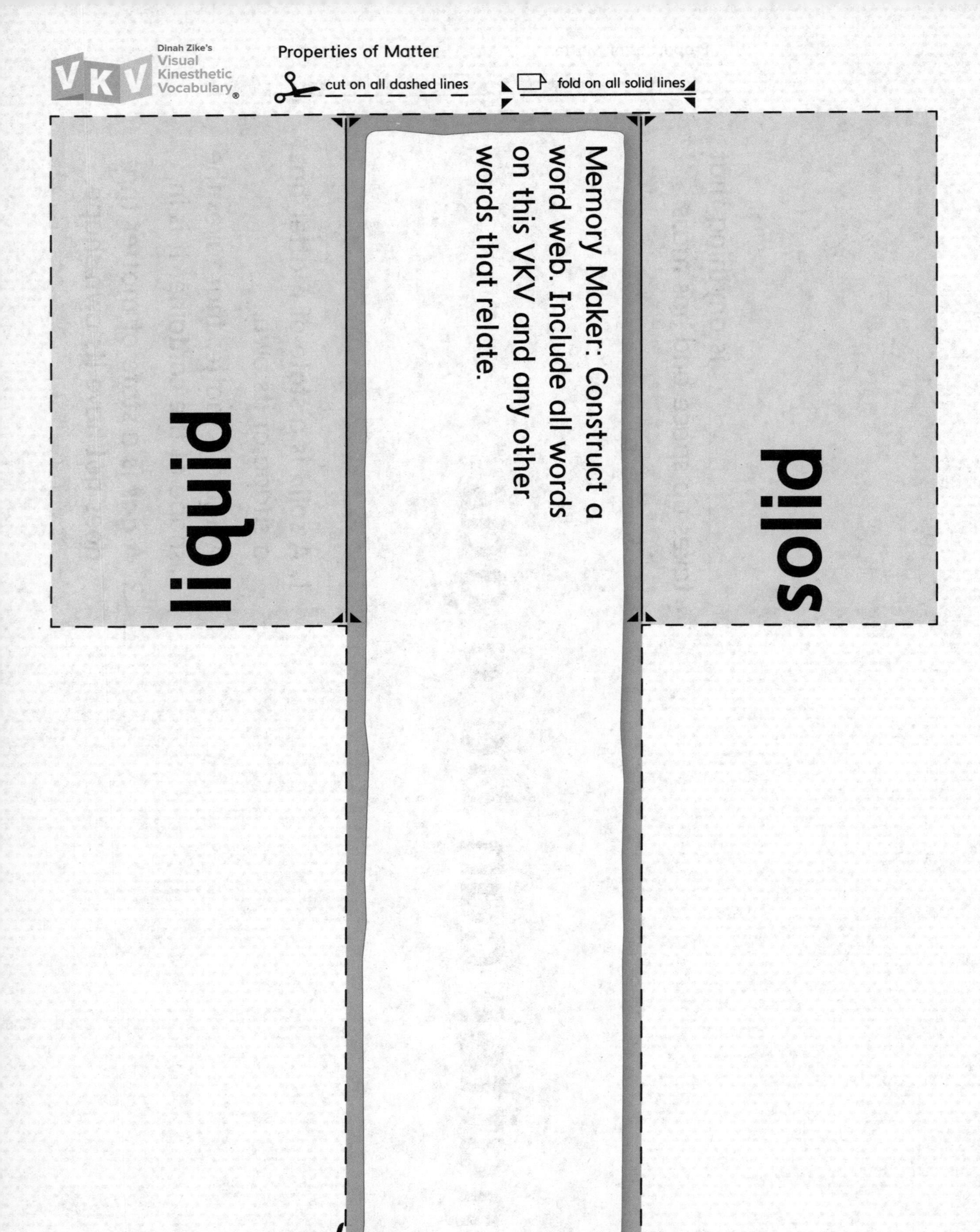

VKV
cut on all dashed lines
fold on all solid lines
liquid
Memory Maker: Construct a word web. Include all words on this VKV and any other words that relate.
solid

cut on all dashed lines

fold on all solid lines

mixture

A ________________ is something made of two or more different things put together.

To ________________ means to change from a solid to a liquid.

To ________________ means to change from a liquid to a solid.

You add heat to ________________ something.

You take away heat to ________________ something.

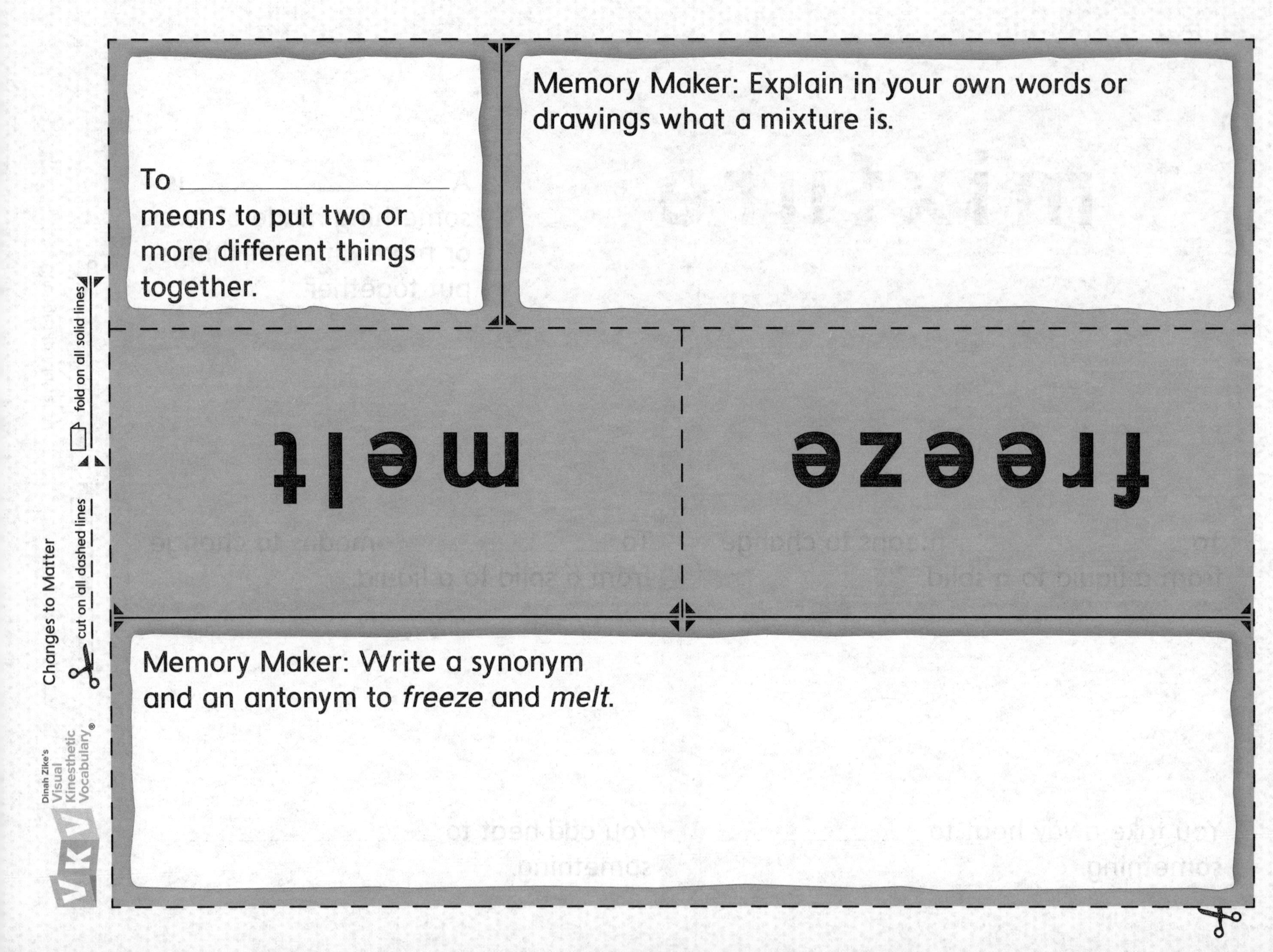
Dinah Zike's
Visual
Kinesthetic
Vocabulary®
VKV
Changes to Matter
cut on all dashed lines
fold on all solid lines
To ________
means to put two or
more different things
together.
Memory Maker: Explain in your own words or
drawings what a mixture is.
melt
freeze
Memory Maker: Write a synonym
and an antonym to *freeze* and *melt*.

cut on all dashed lines

fold on all solid lines

To ______________ something means to change from a gas to a liquid.

condense

1. To **disassemble** something means to take apart.
2. To **assemble** something means to gather into a group.

assemble

To ______________ something means to change it from a liquid to a gas.

evaporate

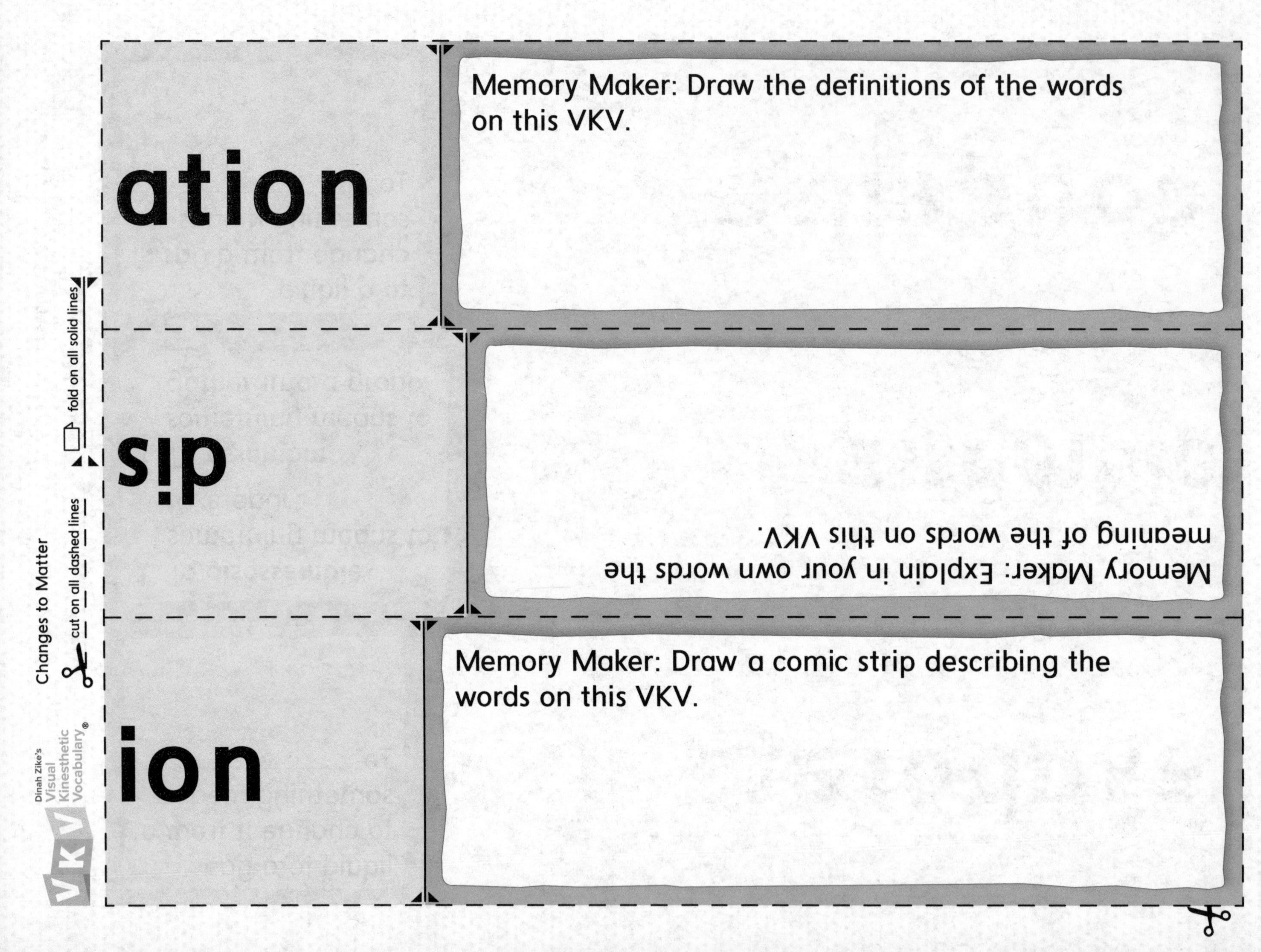
ation
Memory Maker: Draw the definitions of the words on this VKV.
dis
Memory Maker: Explain in your own words the meaning of the words on this VKV.
ion
Memory Maker: Draw a comic strip describing the words on this VKV.
Dinah Zike's Visual Kinesthetic Vocabulary® VKV
Changes to Matter
cut on all dashed lines
fold on all solid lines

VKV Dinah Zike's Visual Kinesthetic Vocabulary®

Changes to Matter

cut on all dashed lines

fold on all solid lines

Memory Maker: Draw a physical change and a chemical change.

physical

A ______ is when something becomes something else.

Changes to Matter

cut on all dashed lines

fold on all solid lines

A ______ change is a change in the way matter looks.

chemical change

A ______ change is when matter changes into different matter.

cut on all dashed lines

fold on all solid lines

Memory Maker: Explain in your own words the terms on this VKV.

digital

A ________ is a tool that measures temperature.

cut on all dashed lines

fold on all solid lines

mercury thermometer

°F
110
100
90
80
70
60
50
40
30
20
10
0

cut on all dashed lines

fold on all solid lines

Memory Maker: Construct a word web. Include all words on this VKV and any other words that relate.

fresh

cut on all dashed lines

fold on all solid lines

salt water

_____ is water that is not salty.

_____ is water with salt in it that is found in ocean.

Earth's Surface

cut on all dashed lines

fold on all solid lines

land

A ______ is a body of water that has land all around it.

A ______ is a small body of fresh water.

A ______ is smaller than a ______.

A ______ is one of the different shapes of Earth's land.

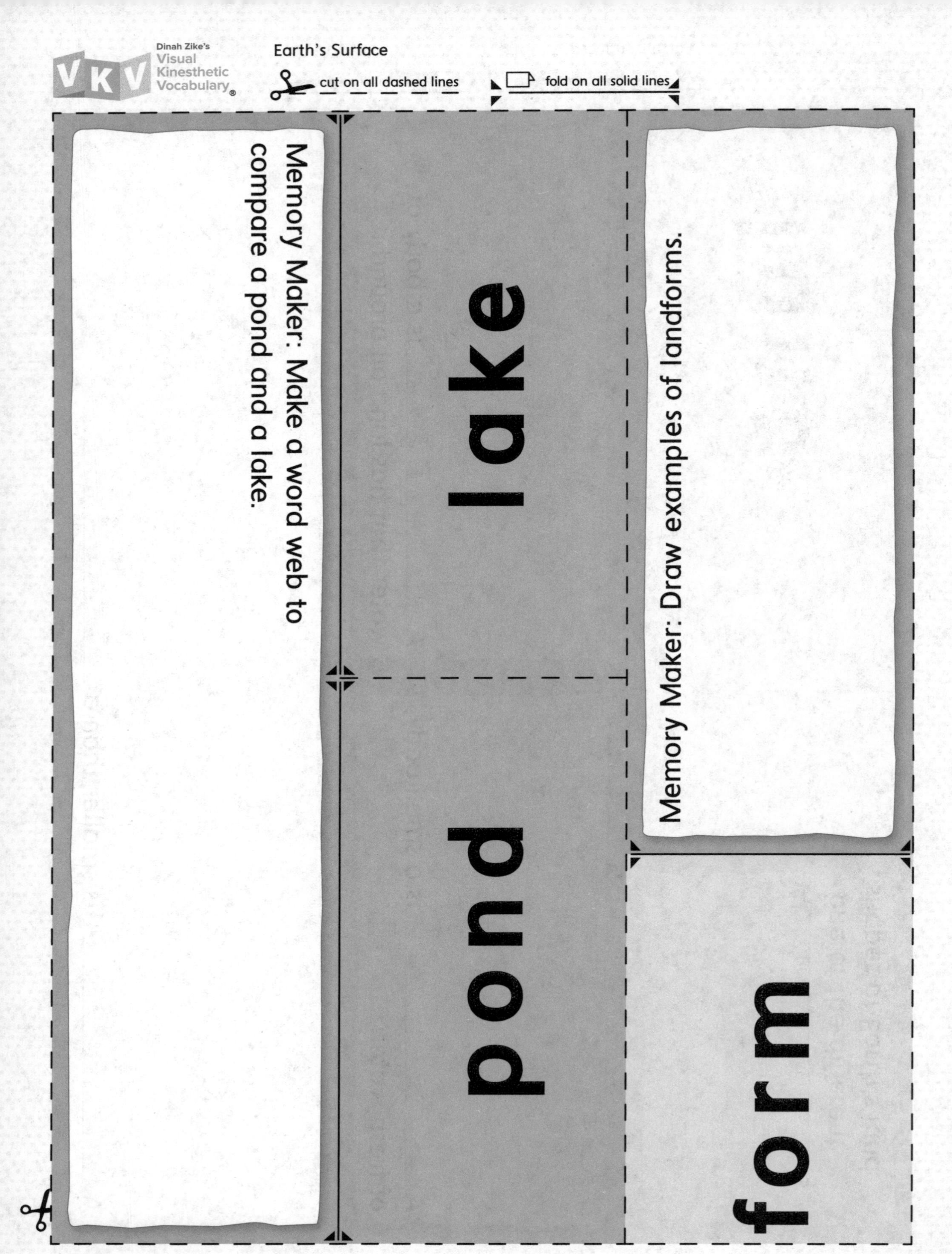
VKV
cut on all dashed lines
fold on all solid lines
Memory Maker: Make a word web to compare a pond and a lake.
lake
pond
Memory Maker: Draw examples of landforms.
form

cut on all dashed lines

fold on all solid lines

An ______ is an area of land surrounded by water.

land

A ______ is the low land between mountains.

If there are many mountains, then there are many ______.

A ______ is land that is very high.

The Rockies are a ______ range.

Dinah Zike's Visual Kinesthetic Vocabulary®

Earth's Surface

 cut on all dashed lines

fold on all solid lines

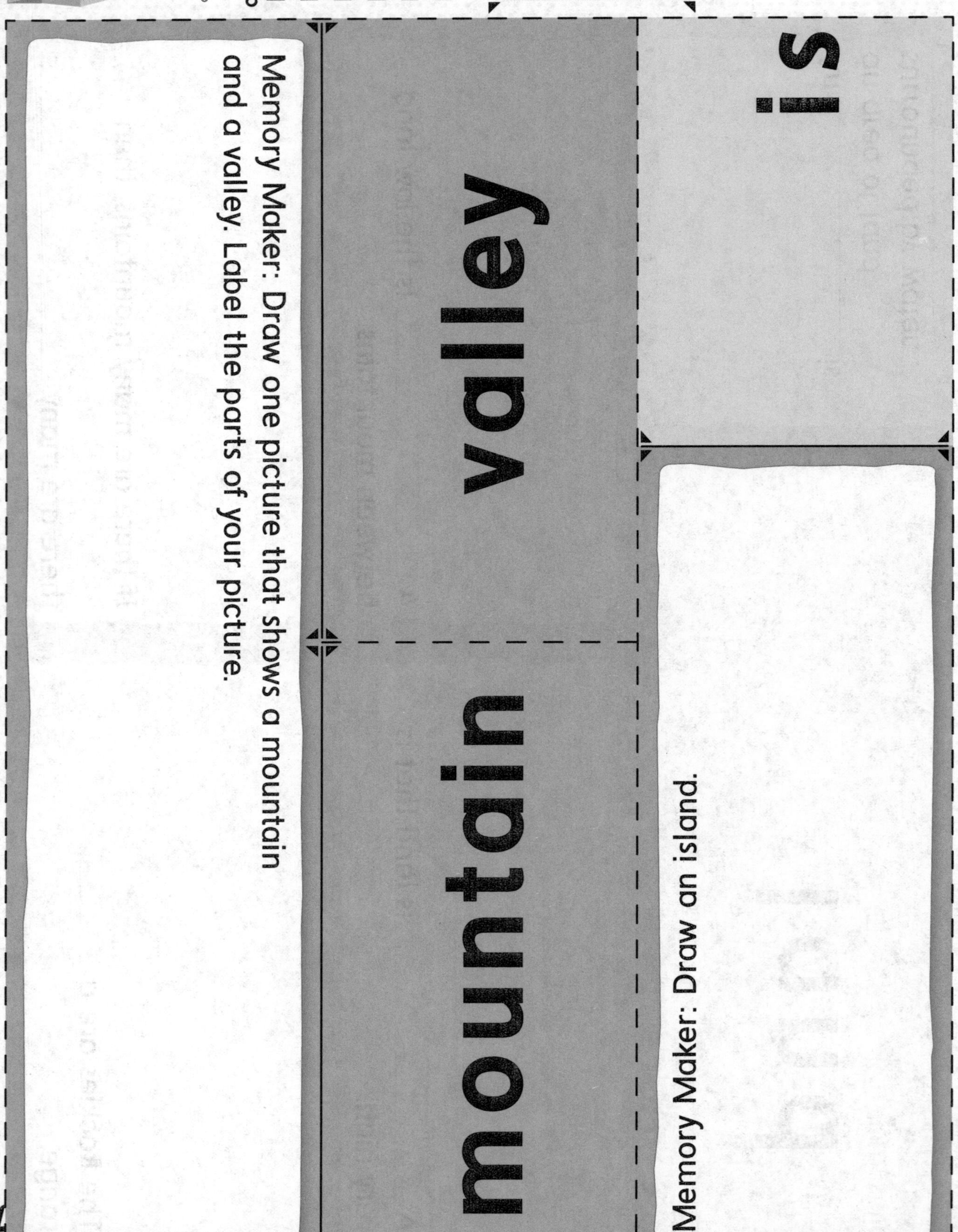
Memory Maker: Draw one picture that shows a mountain and a valley. Label the parts of your picture.
valley
mountain
is
Memory Maker: Draw an island.

cut on all dashed lines

fold on all solid lines

glacier

1. A ______ is a large sheet of ice that moves slowly across the land.
2. **Glacial ice** is frozen, solid water from a glacier.

A ______ is a body of fresh water that flows.

Describe in your own words what a river is.

A ______ is a flowing body of water that is smaller than a river.

Describe in your own words what a stream is.

cut on all dashed lines
fold on all solid lines

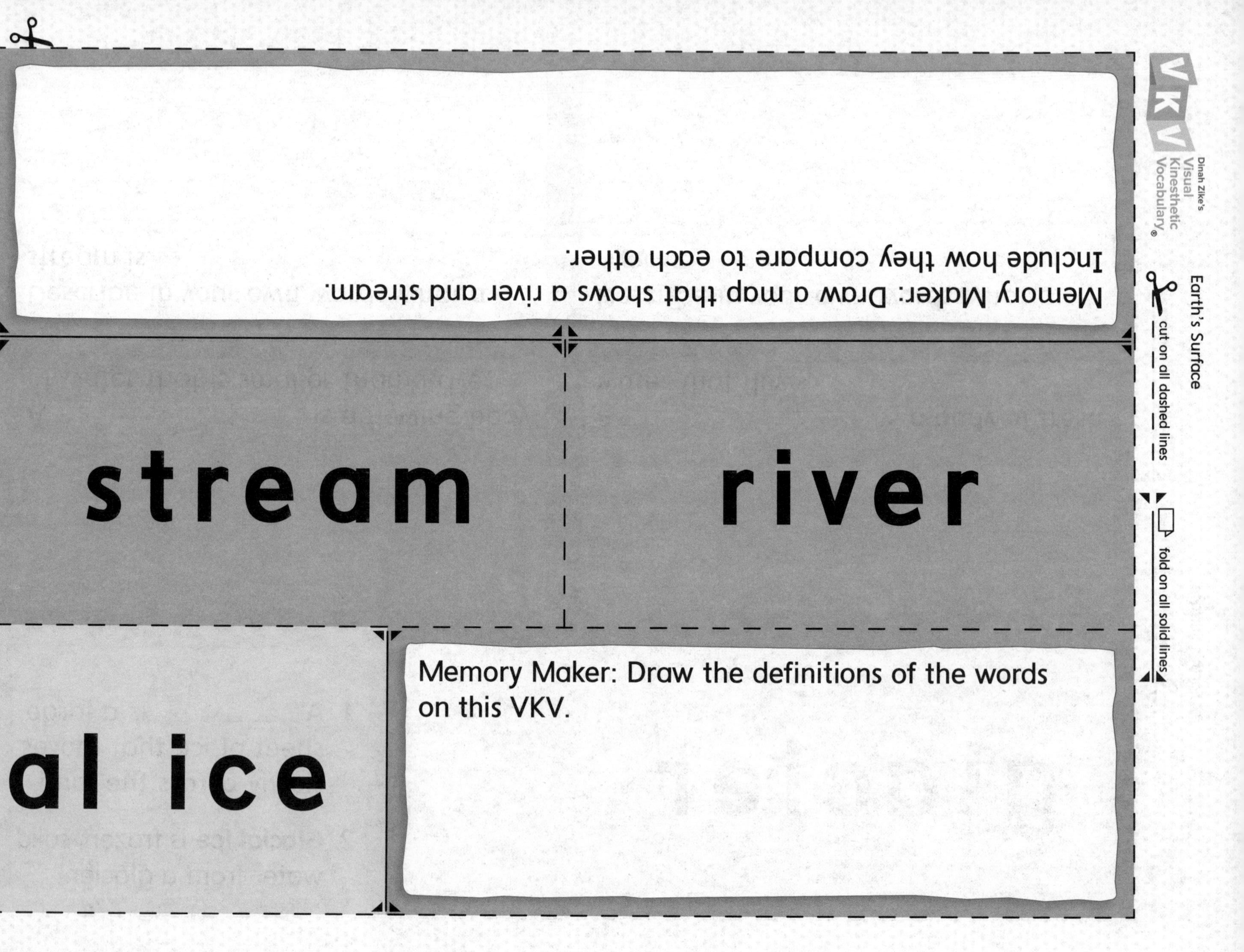
Memory Maker: Draw a map that shows a river and stream.
Include how they compare to each other.
stream
river
al ice
Memory Maker: Draw the definitions of the words on this VKV.

cut on all dashed lines

fold on all solid lines

erosion

_______________ is when rock and soil are moved by wind or water to a new place.

pollution

_______________ is anything that makes air, land, or water dirty.

resource

A _______________ is something that comes from Earth that people use.

Dinah Zike's Visual Kinesthetic Vocabulary®

Earth's Surface Changes

cut on all dashed lines

fold on all solid lines

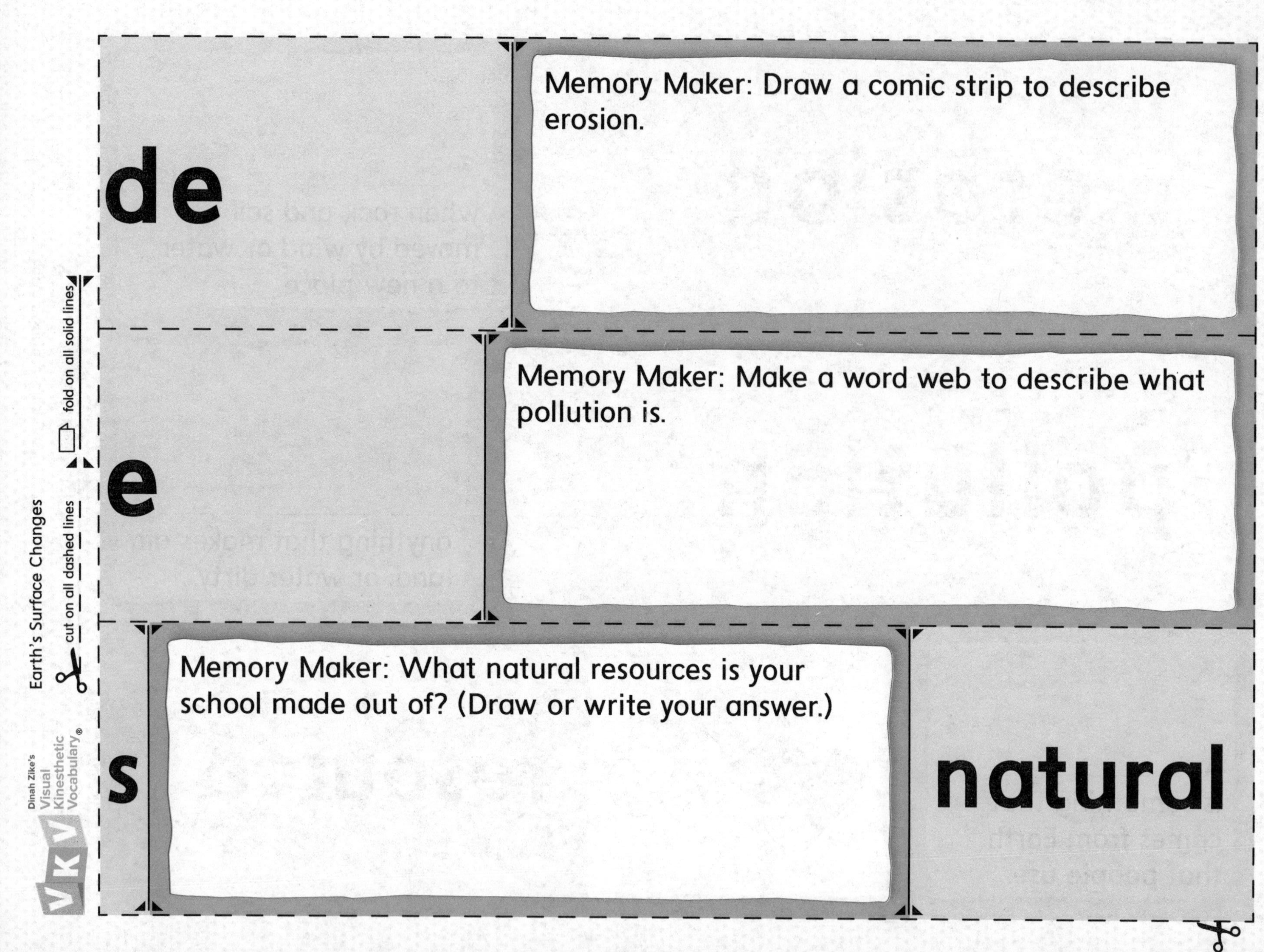

de
Memory Maker: Draw a comic strip to describe erosion.
e
Memory Maker: Make a word web to describe what pollution is.
s
Memory Maker: What natural resources is your school made out of? (Draw or write your answer.)
natural

cut on all dashed lines

fold on all solid lines

1. A **forest** is a habitat where there are many tall trees.
2. The **Arctic** is a very cold place near the North Pole.

1. A ________________ is a place where plants and animals live.
2. A **desert** is a dry habitat that gets very little rain.

grassland habitat

A **grassland** is a large open place with a lot of grass.

cut on all dashed lines

fold on all solid lines

Memory Maker: Draw a picture of each habitat on this VKV.

Arctic

desert

forest

cut on all dashed lines

fold on all solid lines

seed

SEEDS

1. A ______________ is a plant part that can grow into a new plant.
2. A **seedling** is a young plant.

pollen

______________ is a sticky powder inside a flower that helps make seeds.

nutrient

A ______________ is a thing that living things need to grow.

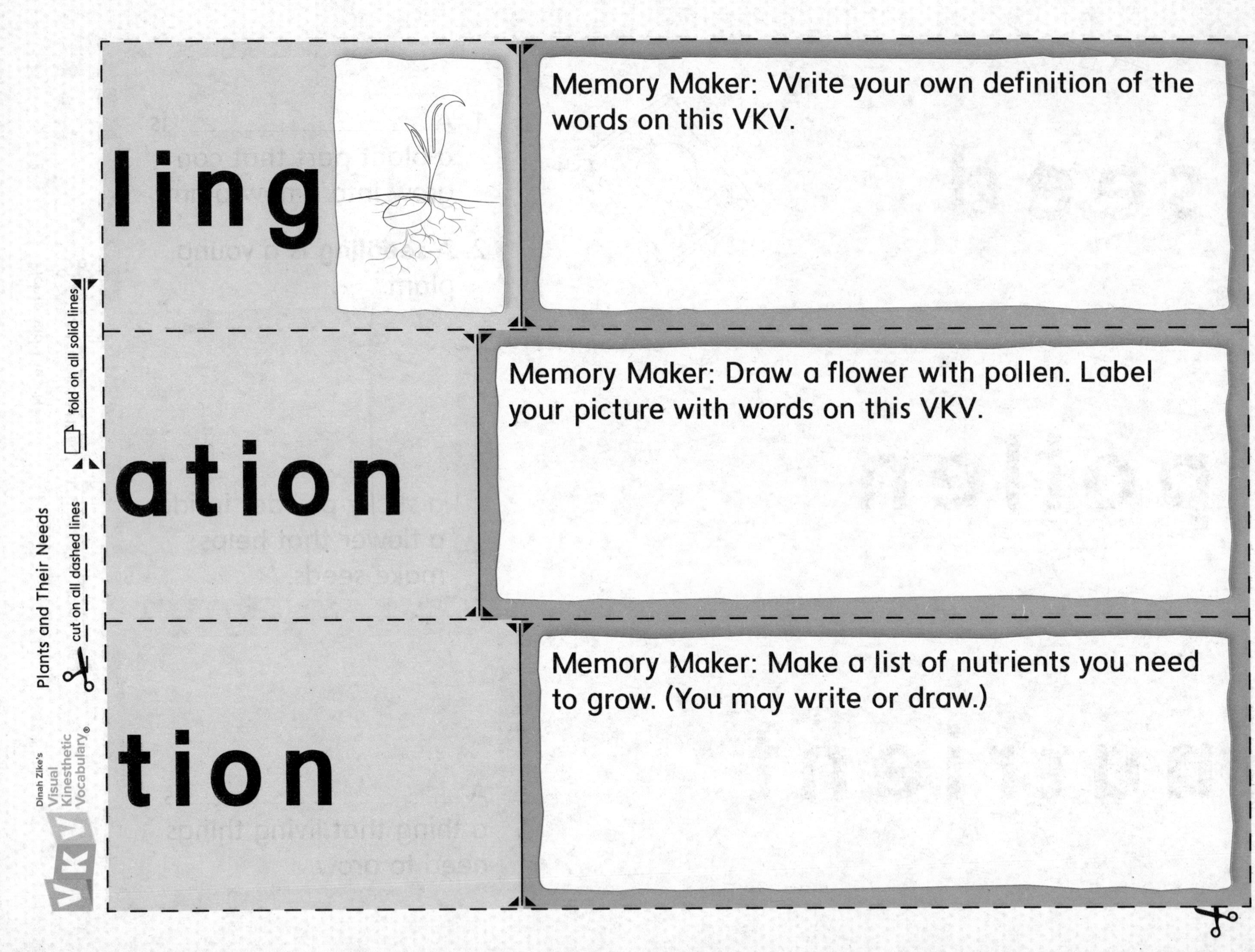
Dinah Zike's
VKV
Visual Kinesthetic Vocabulary®
Plants and Their Needs
cut on all dashed lines
fold on all solid lines
ling
Memory Maker: Write your own definition of the words on this VKV.
ation
Memory Maker: Draw a flower with pollen. Label your picture with words on this VKV.
tion
Memory Maker: Make a list of nutrients you need to grow. (You may write or draw.)
Copyright © McGraw-Hill Education.